ADVANCES IN MOLECULAR AND CELL BIOLOGY

Volume 16 • 1996

CELL ADHESION

ADVANCES IN MOLECULAR AND CELL BIOLOGY

CELL ADHESION

Series Editor: E. EDWARD BITTAR
Department of Physiology
University of Wisconsin

Guest Editor: DAVID R. COLMAN
Brookdale Center for Molecular Biology
The Mount Sinai Medical Center

VOLUME 16 • 1996

JAI PRESS INC.

Greenwich, Connecticut *London, England*

55 Old Post Road No. 2
Greenwich, Connecticut 06836

JAI PRESS LTD.
38 Tavistock Street
Covent Garden
London WC2E 7PB
England

ISBN: 0-7623-0143-0

Manufactured in the United States of America

CONTENTS

LIST OF CONTRIBUTORS

David R. Colman	Brookdale Center for Molecular Biology The Mount Sinai Medical Center
Pamela Cowin	Department of Cell Biology New York University Medical Center
Joseph P. Doyle	Brookdale Center for Molecular Biology The Mount Sinai Medical Center
Donatella D'Urso	Institute Superiori di Sanita Rome
Marie T. Filbin	Department of Biological Sciences Hunter College City University of New York
Gerald B. Grunwald	Department of Pathology, Anatomy, and Cell Biology Thomas Jefferson University
John J. Hemperly	Becton Dickinson Research Center Research Triangle Park, North Carolina
Mark A. Jutila	Veterinary Molecular Biology Montana State University
Eugene E. Marcantonio	College of Physicians and Surgeons Columbia University
Sailaja Puttagunta	Department of Cell Biology New York University Medical Center

Charles L. Rosen	Department of Cell Biology New York University Medical Center
James L. Salzer	Departments of Cell Biology and Neurology New York University Medical Center
Arie F. Struyk	Department of Cell Biology New York University Medical Center
Manhar Wong	Department of Biological Sciences Hunter College City University of New York
Keija Zhang	Department of Biological Sciences Hunter College City University of New York

PREFACE

One prerequisite for the evolution of multicellular organisms was the invention of mechanisms by which cells could adhere to one another. At some point in our history, dividing cells no longer went their separate protozoic ways in the primordial oceans, but instead found that by maintaining an association, by sticking together but not fusing, numerous evolutionary advantages became possible. The subsequent development of specialized tissues and organs depended on the elaboration of incredibly sophisticated, regulatable cell-to-cell adhesion mechanisms which are known to operate in biological processes as diverse as the growth of the embryo, the immune response, the establishment of connections between nerve cells, and arteriosclerosis, to name just a few. Although we can only guess at the ancestral mechanisms that fostered the first primitive intercellular unions, some one billion years ago, we now recognize contemporary molecular "themes" with presumably ancient origins that mediate cell-cell interactions.

The chapters in this book serve as useful, thought-provoking, but not exhaustive, commentaries on contemporary topics within the broad field of cell adhesion. If the reader detects a slight tilt toward those adhesion molecules that function in the nervous system, this is merely a reflection of this editor's interests, biases, and of course, limitations.

David R. Colman
Guest Editor

THE STRUCTURE AND FUNCTION OF INTEGRINS

Eugene E. Marcantonio

I. INTRODUCTION

I will attempt to summarize the experimental data that enlightens us in the structure-function relationship of integrin receptors. I will generalize as much as possible within the diverse structures and growing number of integrin subunits.

Advances in Molecular and Cell Biology,
Volume 16, pages 1-29.

ISBN: 0-7623-0143-0.

II. GENERAL

Integrin receptors are a family of transmembrane glycoproteins that interact with a wide variety of ligands including extracellular matrix glycoproteins, complement, and membrane proteins. These receptors participate in cell-matrix and cell-cell adhesion in a large number of physiologically important processes, including embryologic development and morphogenesis, hemostasis and thrombosis, wound healing, malignant transformation, and leukocyte helper and killer function.

Integrins are noncovalently linked heterodimers with distinct α and β subunits. There are at least eight homologous β subunits and 15 different α subunits, which have been divided into classes on the basis of their component β subunit (Hemler, 1990; Springer, 1990; Hynes, 1992). I have summarized the integrin receptor family in Table 1. The list of ligands is a simplification; in many cases the exact ligand specificity is controversial. (It is beyond the scope of this chapter to discuss these controversies. For a more complete list see Hynes, 1992.) In several cases, it is known that different integrins bind to the same matrix protein via different recognition sequences within the ligand. For example, $\alpha_5\beta_1$ binds to the RGD sequence in fibronectin (FN), while $\alpha_4\beta_1$ binds to the alternatively spliced V25 region of FN. Similarly, $\alpha_1\beta_1$ binds to the E1-4 fragment of laminin, while $\alpha_6\beta_1$ binds to the E8 fragment (see Table 1). Thus, these receptors are a versatile system for recognition of a wide variety of signals.

The integrin α subunits each have a large extracellular domain with three or four putative divalent cation binding domains and in some cases consist of two disulfide-linked polypeptides termed a heavy and light chain (Hynes, 1987). In addition, some but not all of the noncleaved α subunits have a large extra domain, also called an inserted (I) domain, within the amino terminal portion of the molecule (e.g., Takada and Hemler, 1989; Corbi et al., 1988).

All integrin subunits contain a single transmembrane segment followed by a carboxyl terminal cytoplasmic domain of varying length. Within the β_1 cytoplasmic domain, there is a site for tyrosine phosphorylation in transformed cells (Tamkun et al., 1986; Hirst et al., 1986) and this domain is highly conserved in both vertebrates and invertebrates (DeSimone and Hynes, 1988; Marcantonio and Hynes, 1988). These domains are believed to be involved in the interaction of integrins with the cytoskeleton since deletions and

Table 1. The Integrin Receptor Family

Subunits		*Ligands and Counterreceptors*
β_1	α_1	Collagens (IV>I), laminins (E1-4)
	α_2	Collagens (I>IV), laminins
	α_3	Fibronectin, collagens, laminins (E8)
	α_4	Fibronectin (V25), VCAM-1
	α_5	Fibronectin (RGD)
	α_6	Laminins (E8)
	α_7	Laminins
	α_8	Fibronectin
	α_v	Fibronectin (RGD)
β_2	α_{lfa-1}	ICAM-1, ICAM-2
	α_{mac-1}	C3bi, fibrinogen, factor X, ICAM-1
	α_{p150}	Fibrinogen
β_3	α_{IIb}	Fibrinogen, fibronectin, von Willebrand factor, vitronectin, thrombospondin
	α_v	Fibrinogen, fibronectin, von Willebrand factor, vitronectin, thrombospondin, osteopontin
β_4	α_6	Laminin
β_5	α_v	Vitronectin
β_6	α_v	Fibronectin
β_7	α_4	Fibronectin (V25), VCAM-1
β_8	α_v	Unknown

certain point mutations of the β_1 cytoplasmic domain block the ability of integrins to organize into focal adhesion plaques (Solowska et al., 1989; Hayashi et al., 1990; Marcantonio et al., 1990; Reszka et al., 1992).

The general feature of all integrin receptors is that they provide a transmembrane linkage between the ECM and the cytoplasm, and specifically the cytoskeleton. In addition, these receptors are critical in the cellular response to adhesion, and for recognition of signals provided by the matrix. I will now discuss in detail the structure and relationship of individual domains of integrins within the context of their overall function, beginning with ECM binding and progressing to the inside of the cell.

A. Ligand Binding Properties

Both subunits contribute to specificity. Initially, it appeared that the α subunits within a β subunit class were the primary determinants of specificity. However, the discovery that α subunits, particularly α_v, could associate with more then one β subunit has changed this notion. Since α_v can have a different binding specificity when combined with β_1, β_3, β_5, or β_6, it is obvious that the individual combinations of α/β subunits determine ligand specificity (Busk et al., 1992; Smith and Cheresh, 1990; Vogel et al., 1990).

In addition, there are cell type specific factors that influence specificity as well. This phenomenon has led to some inconsistencies in the literature over which integrins bind which ligands. There are numerous examples, but the most studied has been the integrin $\alpha_2\beta_1$. This integrin is a receptor for collagen, but not laminin on platelets, fibroblasts, and certain lymphocyte lines, while binding both collagen and laminin on endothelial cells (Elices and Hemler, 1989; Kirchhofer et al., 1990). The molecular basis for this cell type difference can be considered in three mechanisms. One obvious possibility would be alternative splicing of either the α or β subunit in a cell-type specific manner. There is indeed evidence for such a phenomenon. However, for the $\alpha_2\beta_1$ subunit, the only evidence is that there can be alternative splicing within the β_1 cytoplasmic domain. There is evidence for alternative exon utilization within the drosophilia β subunit in a region thought to be important for ligand binding, but not for mammalian integrins (Zusman et al., 1993).

There are several reasons to suggest that alternative splicing is not the mechanism of cell type specific ligand specificity. First of all, the binding characteristics appear to be true for virtually all of the $\alpha_2\beta_1$ receptors in a certain cell type. Thus, if alternative splicing is the mechanism, it must be absolute. There must be exclusive utilization of the same sequence in fibroblasts and platelets, while the alternative sequence utilized in endothelial cells. For the alternative splicing of the β_1 cytoplasmic domain, there is utilization of the same form in fibroblasts and endothelial cells (β_{1a}; Balzac et al., 1993). Thus, it is extremely unlikely that alternative β_1 cytoplasmic domains play a role in cell type specific differences in ligand binding. I will return to the alternative splicing of cytoplasmic domains during the discussion of their structure.

Experiments performed in the laboratory of Martin Hemler with $\alpha_2\beta_1$ strongly suggest that there are cell type factors which determine an activation state of an integrin receptor, which are independent of alternative splicing. These investigators have shown that the heterologous expression of an α subunit cDNA in different cell types can lead to different ligand specificity (Chan and Hemler, 1993). Thus, alternative splicing of the α subunit as a primary mechanism for cell type ligand specificity appears to be ruled out. The experiments did not rule out alternative splicing of the β subunit, except to demonstrate that the authentic β_1 cytoplasmic domain was present in both cell types. In fact, much of the experiments done to date suggest that the β subunit is the subunit affected by cell type dependent modification. Chan and Hemler (1993) showed that the $\alpha_2\beta_1$ integrin was in an inactive state on some cells, a collagen binding state on others, and a collagen and laminin binding state in others. This latter state would be consider the most active, while the others less active. Using anti-β_1 antibodies, the less active forms could be made more active, even in a detergent extract of the cells. Thus, the simplest explanation for these results would be that the integrin conformation could be modulated by β_1 antibodies, and this conformation was what determines the cell type specific ligand specificity.

If this is the case, then what determines the conformation differences between cell types? One needs to rule out the role of the β subunit in alternative splicing, which could be done via heterologous expression as has been done for α_2. The results of one study (Hayashi et al., 1990) suggest that alternative splicing of the β subunit is not responsible for the cell type differences. These authors expressed the avian β_1 in a mouse 3T3 cell. This heterologous expression led to a difference in the ligand specificity of the mouse α subunits. The parental 3T3 line does not bind laminin well, despite the presence of $\alpha_3\beta_1$, a known laminin receptor. When the avian β_1 is introduced into these cells, they now bind to laminin, and this adhesion is entirely chicken-specific. Thus, there must be a feature of the chicken β_1 that allows for laminin binding conformation. Although it is possible that the chicken β_1 has an alternative exon from the mouse, this seems unlikely. However, since the genomic structure of the β_1 gene has been only partially characterized (Lanza et al., 1990; Altruda et al., 1990), it remains to be seen if there are any alternative exons in the extracellular domains to be differentially utilized.

Two other possibilities remain. One is that the integrin subunits undergo differential post-translation modifications which lead to changes in ligand binding specificity via conformation changes. One modification to be considered is glycosylation. This post-translation modification is particularly feasible. This process is often cell-type specific, and can be absolute from cell to cell. However, while it is likely that there is heterogeneous glycosylation of integrins in various cell types, there is little data concerning the functional consequences.

Another post-translational modification to be considered is phosphorylation of the cytoplasmic domains of either subunit. While this modification clearly occurs in many cells (Chatila et al., 1989; Buyon et al., 1990; Horvath et al., 1990), there are several reasons to doubt the functional consequences of phosphorylation. First, as discussed previously, since the differences between cells seem to involve virtually all of the receptor molecules in a cell, the stoichiometry of the phosphorylation would be expected to be high. However, studies suggest that this stoichiometry is low (Hillery et al., 1991; Chatila et al., 1989). Secondly, mutagenesis of the cytoplasmic domain to remove sites of phosphorylation in the β_2 integrin subunit had no change in the activation of this integrin (Hibbs et al., 1991). In fact, mutant versions of this subunit that could not be phosphorylated on serine (as is thought to happen with protein kinase C; PKC were activated by treatment of transfected cells with phorbol esters (Hibbs et al., 1991). Thus, activation of integrins via phosphorylation is very unlikely. This type of modulation might be important in rapidly reversible changes, while the more deliberate differences between cells during differentiation (as discussed later in the chapter) would be difficult to maintain via phosphorylation. It seems likely to me that phosphorylation can play a role in rapidly down regulating integrin function, possibly in migration.

The last possibility I will discuss is that of an additional component(s) that is responsible for the changes in ligand binding from cell to cell and/or for activation of integrin binding. There is some evidence in support of this hypothesis and I personally believe this is the mostly likely mechanism. The evidence comes from the studies of inside-out signaling or activation of integrins. In particular, there are four separate examples of lipids that modulate integrin function. In neutrophils, a neutral lipid has recently been isolated that is required for activation of integrin function. This lipid, called Integrin Modulating Factor, or IMF, is generated upon activation of neutrophils with F-met-leu-phe (Hermanowski-Vosatka et al., 1992). This lipid increases the binding of a number of ligands to Mac-1, both in intact cells and in purified receptor preparations. Furthermore, the acidic lipids phosphatic acid (PA) and lysophosphatic acid (LPA) have been shown to modulate the binding of purified $\alpha_{IIb}\beta_3$ to fibrinogen and appear to work best when incorporated into lipisomes with the purified receptor (Smyth et al., 1992). Both IMF and LPA have been shown to bind directly to their respective integrins. In addition, Conforti and colleagues (1990) have shown that changes in the lipid environment can produce subsequent changes in the binding characteristics and physical conformation of $\alpha_v\beta_3$. Earlier experiments (Cheresh et al., 1987) had shown that this integrin binds gangliosides in certain cell types, and that this binding was required for function. Thus, a diverse number of lipids appear to modulate integrin function in a number of cells. Since these lipids can perform this modulation in purified receptor preparations, they must be directly affecting the conformation of integrins, rather than acting as intermediates in signalling pathways. I will return to the possible role of lipids during the discussion of the transmembrane regions.

In addition, recently an additional factor, called cell adhesion regulator, has been isolated by expression cloning, looking for genes that up-regulated cell adhesion (Pullman and Bodmer, 1992). This protein has a tyrosine that may be phosphorylated for function, and may be an intermediate in signalling pathways which lead to integrin activation.

Many of these issues of mechanism of activation in circulating cells also need to be addressed in the studies of control of integrin functions in differentiation, which may be more stable changes. There is now considerable evidence that the main regulatory mechanism by which cells modulate integrin function is inactivation on the cell surface, not by modulation of the expression level. For example, in keratinocytes, the adhesion to FN, mediated by $\alpha_5\beta_1$, is quite vigorous as the cells remain as basal cells. However, as these cells differentiate, this adhesion is lost, while similar cell surface levels of $\alpha_5\beta_1$ persist in the cell (Adams and Watt, 1990). Thus, an inactivation of the $\alpha_5\beta_1$ appears to have occurred. Similarly, Neugebauer and Reichardt (1991) have shown that retinal neurons lose the ability to adhere to laminin upon maturation in the embryo. This adhesion is dependent on $\alpha_6\beta_1$, whose cell surface levels remain constant throughout the maturation process. Furthermore, when activating β_1 antibodies were added, the mature neurons now regained the ability to bind laminin via $\alpha_6\beta_1$, suggesting that this receptor had been in an inactive state. It seems likely that multiple mechanisms are used by cells for both rapidly reversible and long term modulation of integrins, and that these mechanisms will have different wrinkles depending on the integrin, the cell type, and the process involved.

B. Ligand Binding Domains

The extracellular domains of both α and β subunits have been extensively studied to elucidate the segments responsible for ligand binding. Although much information exists in regards to ligand binding domains, there are major gaps in our knowledge. A model of the domain structures of two kinds of α subunits and the β subunit is shown in Figure 1. The first type of α subunit shown has a post-translational intrachain cleavage and four discrete metal binding domains, each contained with a repeat. The other form has an I or inserted domain which interrupts the modular structure, and has three discrete divalent binding sites within repeats 5-8 (but see also below). Since virtually all integrin ligand binding is divalent cation dependent, I will begin with a discussion of putative cation binding domains and their role in ligand binding. It would be impossible to separate the two subjects, since it is now clear that all of the putative ligand binding domains in the α and β subunits are also potential cation binding domains.

The majority of studies concerning cation binding domains have been concentrated on the integrin $\alpha_{IIb}\beta_3$. Brass and Shattil (1984) showed that the majority of Ca^{2+} binding sites on the surface of resting platelets are lost in Glanzmann's thrombasthenia platelets, an inherited deficiency in $\alpha_{IIb}\beta_3$. There are clearly two high affinity sites (Kd=9nM) and several low affinity sites (Kd=400nM). Initially, after the first integrin α subunits were cloned and sequenced, it was thought that the major divalent binding sites were the four domains found in α_{IIb} (Ponz et al., 1987) and α_5 (Argraves et al., 1987), which have a sequence homologous to the consensus metal binding domain sequences found in calmodulin (Kretsinger, 1980). Subsequently, the situation has been found to be much more complex than previously considered.

It remains unclear what is the precise role of divalent cation binding regions in ligand recognition. A number of studies have shown that specific cations are preferentially utilized. For example, manganese has been shown in several instances to up-regulate the affinity of integrins for their ligands and to induce the appearance of activation-specific epitopes (Gailit and Ruoslahti, 1988; Dransfield et al., 1992).

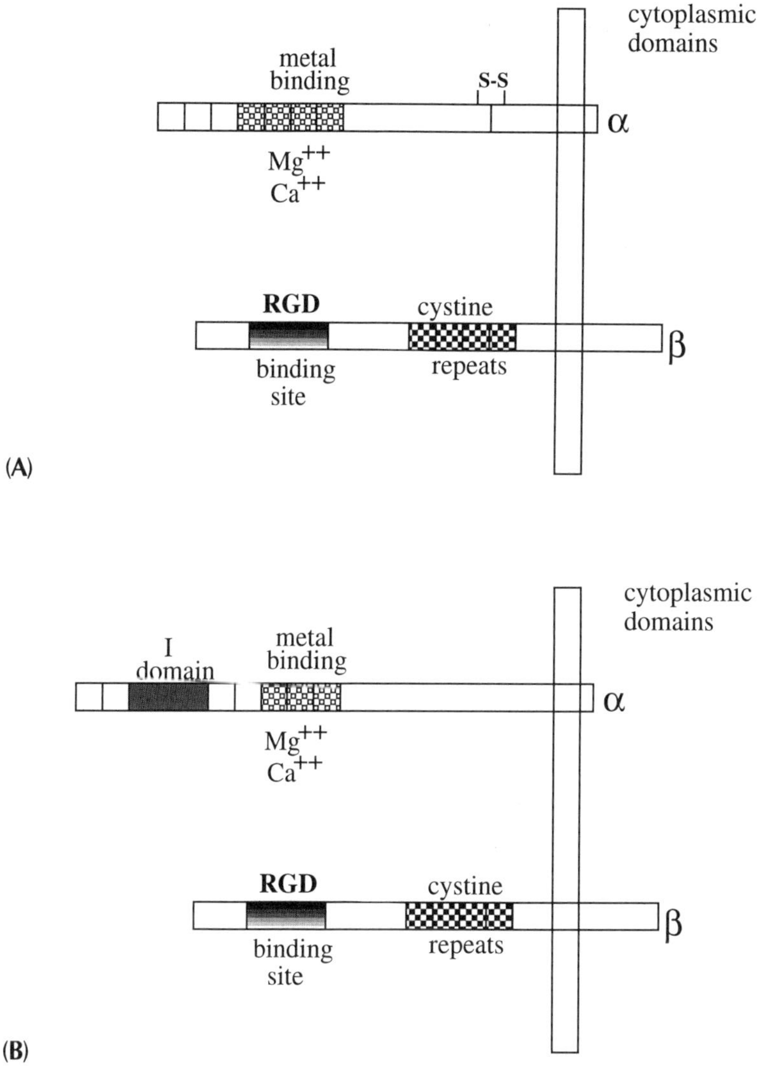

Figure 1. Diagram depicting the two main classes of integrin heterodimers. **(A)** A heterodimer between β and a cleaved α subunit is shown. Note the intrachain disulfide bond which connects the heavy and light chains of this type of α subunit. There are four metal binding domains contained in repeats 4-7. **(B)** A heterodimer between β and an I domain containing (or non-cleaved) α subunit is shown. The repeating structure is interrupted by a large inserted (I) domain. There are only three of the corresponding metal binding domains of the cleaved α subunits, in repeats 5-7.

The motifs believed to be involved in cation binding are present in extracellular portion of alpha subunits. These domains are contained within the repeats 4-8, and have a consensus that is similar to a portion of an E-F hand structure. The intriguing feature of these metal binding domains is their lack of the complete number of coordination oxygens for the binding of cations as based on the structure of calmodulin. This information must be tempered with the realization that the divalent cation concentration in the extracellular space is in the millimolar range, which may allow binding even without all of the coordinating oxygen residues being present. Thus, there may be bound divalent cation in all metal binding domains of the integrins.

Recent structure-function studies, primarily of the platelet integrin α_{IIb}/β_3, have provided insight into the role of these metal binding domains in the mechanism of integrin-ligand interactions. As mentioned previously, the structure of these domains are such that a critical oxygen residue is not present to complete the E-F hand structure, so characteristic of divalent cation binding. Initially, Corbi and colleagues (1988) proposed that this oxygen residue could be provided by the ligand. Based on this hypothesis, D'Souza and colleagues (1991) have synthesized a peptide derived from a metal binding domain (repeat five) of α_{IIb}, and tested this peptide for its effect on ligand binding. The wild type peptide blocked the binding of fibrinogen to α_{IIb}/β_3 while a peptide with a single substitution (asp to glu) had no effect on binding. These results support the hypothesis that the ligand could provide the final coordination oxygen around a metal binding motif, and this interaction could form the basis of many integrin ligand interactions. This mechanism helps explain the pivotal role of the RGD recognition sequence in FN and other integrin ligands. This sequence is not active when the aspartate is substituted for by a glutamate, which would be very much compatible with the aspartate residue coordinating its oxygen side chain around a divalent cation, thus mimicking ligand binding. The use of these metal binding domain peptides gave us hope that it would be easy to design peptide inhibitors of integrins based on their metal binding domain sequences. We have tried the corresponding peptides from the human α_1 sequence (Briesewitz et al., 1993a), with no effect on ligand binding (unpublished observations). Thus, the universal role of these domains in integrin ligand binding remains unclear.

Furthermore, recent work has shown that these divalent cation binding regions are not the whole story. A number of experiments strongly suggest that another domain, the I domain or inserted domain (also known as an A domain), which is present in a subset of α subunits, is involved in ligand recognition. There are five α subunits which belong to this distinct subset of integrins. These α subunits are α_1, α_2, α_{lfa}, α_{mac-1}, and α_{p150}. All contain an approximately 210 amino acids domain, which is inserted between repeats 3 and 4 in the modular structure common to all α subunits. There are several features of this group of α subunits that are unique. First, and perhaps foremost, is the intriguing feature of a preference for Mg^{2+}, for which Ca^{2+} cannot substitute (e.g., Rothleim and Springer, 1986; Staatz et al., 1989). Indeed, Grzesiak and colleagues (1992) have shown that high levels of Ca^{2+} can inhibit the function of $\alpha_2\beta_1$, providing for another mechanism of regulation of these integrins. The role of the I domain in binding divalent cations has been demonstrated very recently. Expression of this segment as a bacterial fusion protein led to the recovery of a protein segment capable of binding several divalent cations (Michishita et al., 1993). Interestingly, this region preferred Mn^{2+}, then Mg^{2+} with Ca^{2+} bound less well. This preference exactly parallels that of all of the I domain containing integrins. Furthermore, mutation of an aspartate residue within the I domain of full length Mac1 led to a receptor with impaired ligand binding properties.

The I domain has been implicated in collagen binding of integrins, based on sequence homology of this region with collagen binding domains of von Willebrand factor (Pytela, 1988). In addition, Diamond and colleagues (1993) have shown that a large number of α subunit blocking antibodies bind to the I domain of α_{mac-1}. Together all of these findings suggest a significant role for the I domain in ligand binding. It remains to be seen if a corresponding region of the other α subunits is involved in ligand binding.

Insights into the role of the β subunit in ligand binding have come from a number of sources. Initially, two groups successfully used photoaffinity crosslinking data to demonstrate the regions of α and β subunits which contact RGD peptides during binding (Smith and Cheresh, 1988; D'Sousa et al., 1988). The identified region in the β_3 subunit is residues 109-170 in the amino terminal region. The first 20 amino acids of this region are particularly interesting because of the high degree of conservation between β subunits. Furthermore, this region also may contribute to divalent cation binding. Recent studies (Loftus et al., 1990) have shown that mutations either naturally occurring or introduced *in vitro* in this region, which would destroy the β_3 cation binding, have dramatic effects on ligand binding as well. Furthermore, Takada and colleagues (1992) have mutated *in vitro* the corresponding aspartate residue in the β_1 subunit, which perturbs ligand binding. Lastly, epitope mapping of blocking antibodies to the β_1 subunit show that the region bound is very close to the putative RGD binding domain (Shih et al., 1993). This domain is very highly conserved among all β subunits, despite the fact that many of the β subunits do not bind RGD or the many RGD dependent ligands. Thus, it seems likely that the RGD binding domain within the β subunit plays a pivotal role in ligand binding.

There is a series of experiments that suggests a model for integrin ligand binding. Thus, the model predicts that the ligand, α subunit, and β subunit all contribute to divalent cation binding, which stabilizes the whole complex. Thus, divalent cation binding is not only required for binding, but may be central to the final ligand interaction with the receptor. However, it is clear that additional sequences both in the ligand and receptor are important in recognition, and these interactions may not involve cation binding.

C. Subunit Association

Several studies have shown that association of the native (intact) α and β subunits of integrins is required for expression on the cell surface (Ho and Springer, 1983; Duperray et al., 1989; Rosa and McEver, 1989). Furthermore, β_3 subunits will not be transported to the surface in the absence of the α subunit (Cheresh and Spiro, 1987). This requirement for dimerization helps explain the lack of α and β subunits in patients with genetic alteration in one of the two subunits as seen in Glanzmann's thrombasthenia and in leukocyte adhesion deficiency (LAD). In the latter disease, defective β_2 subunits can be synthesized, that will not dimerize with α subunits, leading to a profound deficiency of all three β_2 heterodimers (Kishomoto et al., 1989). If either α or β is missing, the other subunit is retained intracellularly, presumably in the ER. Thus, it seems that both the integrin α and β subunits have "signals" that mediate ER retention or degradation. Analysis of the assembly of the T-cell receptor subunits have shown that sequences involved in ER retention or degradation can also be critical for subunit assembly (Bonifacino et al., 1991a, 1991b; Blumberg et al., 1990; Alcover et al., 1990). Thus, one mechanism to escape ER retention would be for the retention signal to be hidden by association with sequences on another subunit. Therefore, definition of the ER retention signals within the α and β subunits may help identify key regions associated in dimerization.

Investigations of the domains involved in subunit association have focused on the production of secreted forms of integrins. Several labs have produced soluble forms of integrins, using site directed mutagenesis (Dana et al., 1991; Frachet et al., 1992; Briesewitz et al., 1993a; Bennett et al., 1993). In all cases, deletion of the transmembrane and cytoplasmic domains of α and β led to the secretion of functional heterodimers. For both the α_1 (Briesewitz et al., 1993a) and α_{IIb} (Bennett et al., 1993) subunits the secretion is dependent on dimerization, that is, α subunit alone is not secreted. Thus, the ER retention signal persists in the truncated forms of α subunits. Furthermore, using deletion analysis, we have localized the retention signal to the amino terminal half of the extracellular domain of α_1. For the α subunit one could conclude that this region is critical for dimerization since assembly is required to avoid ER retention.

The situation is quite different for the β subunits. Deletion of the transmembrane and cytoplasmic domains of the β subunit leads to its secretion with a similar transit time as when assembling with the secreted α subunit. Therefore, we may conclude that the ER retention signal has been lost. One possibility is that the retention signal is contained within either the transmembrane or cytoplasmic domain. However, experiments with reporter constructs have recently ruled out both of these regions. Laflamme and colleagues (1992) have shown that the β_1 cytoplasmic domain, when attached to a reporter membrane protein, does not lead to ER retention. Furthermore, Geiger and colleagues (1992) have shown that a construct containing the extracellular portion of N-cadherin fused to the integrin β_1 transmembrane and cytoplasmic domain is efficiently transported to the plasma membrane. By these analyses the cytoplasmic, transmembrane, and extracellular domains as individual regions cannot mediate ER retention. However, it is clear that the intact β subunits have such a mechanism that does not function well in reporter constructs.

The ability of truncated integrins to assemble into functional heterodimers have led some to conclude that there is no role for the transmembrane and cytoplasmic domains in the dimerization of native integrin subunits. There are some unanswered questions concerning integrin assembly. Recently, Briesewitz and colleagues (1993a) have produced soluble, truncated versions of the integrin $\alpha_1\beta_1$, and compared the assembly of the truncated heterodimers with native heterodimers in 3T3 cells. Despite the fact that the intact human α_1 can associate with native mouse β_1 (Briesewitz et al., 1993a), and the intact avian β_1 can interact with the native mouse α subunits (Solowska et al., 1989; Hayashi et al., 1990), the truncated forms of the human α_1 and the avian β_1 do not interact with the endogenous subunits, but instead preferentially interact with each other. Since the dimerization of truncated α and β is independent intracellularly from the native α and β, one possibility is that the transmembrane and cytoplasmic regions of α and β may play a role in dimerization of *intact* integrin subunits. An alternative hypothesis would be that the segregation of soluble versus membrane heterodimer formation is due to differences in the assembly of secreted versus membrane proteins in general. We have designed experiments to test whether the specific α and β transmembrane and cytoplasmic segments are required, or whether nonspecific segments are sufficient.

These experiments involve replacement of the native transmembrane and cytoplasmic domains of α and β subunits with their respective partner in a reciprocal fashion, or with interleukin receptor transmembrane domains (Briesewitz et al., 1995). All transmembrane chimeric versions of α_1 with either exogenous (not native) transmembrane domains were retained in 3T3 cells because of their inability to dimerize with the endogenous β subunits. Similarly, the β subunits containing exogenous transmembrane domains were retained in the cells, due to their inability to dimerize with endogenous α subunits. Interestingly, these chimeric β subunits have α transmembrane and cytoplasmic segments, which we have shown do not mediate α retention. Presumably, when the β extracellular domain is membrane bound its "retention signal" functions, while when in a soluble form, this signal does not function. These experiments do not explain the mechanism of exogenous versus endogenous dimerization. To address this question, chimeric α and β subunits were introduced into the same cell line, which leads to *selective* dimerization of these chimera, and a lack of dimerization with endogenous integrins, regardless of the nature of the transmembrane domain. However, the α and β cytoplasmic domains *must* be on opposing chains for this selective dimerization. These chimeric subunits will not dimerize with soluble forms, and deletion of the conserved GFFKR in the α cytoplasmic domains prevents both endogenous and exogenous dimerization. These results strongly suggest that the proximal portions of the α and β cytoplasmic domains are associated, and that this association is required for normal assembly of integrins. I will return to this issue in the description of a model of integrin function (*vide infra*).

D. Transmembrane Domains

The most obvious feature of the transmembrane domains of both α and β subunits is their striking sequence homology. For example, there are a number of residues in the α and β transmembrane domains which are invariant, or at least only have very rare conservative substitutions. These conserved residues are involved in the maintenance of a precise α helical structure. This striking conservation predicts an important general receptor function. As described previously, swapping of foreign transmembrane domains into integrins does not affect heterodimer assembly as long as the α and β cytoplasmic domains are opposed. However, these chimeric receptors have a disruption in post ligand binding events, particularly in formation of focal adhesions, which demonstrates the role of the native transmembrane domains in the flow of information intramolecularly through integrins. Although the precise mechanism of the role of transmembrane domains in integrin function remains unclear, it is likely to be an important one. Based on this assumption, it is understandable that changes in lipid composition could have dramatic effects on integrin function, as described above.

E. Post-ligand Binding Events

After cells attach to a matrix via integrin receptors, there are a series of possible post-ligand binding events. In fibroblasts, binding to most ECM molecules leads to spreading of the cell with organization of the cytoskeleton. Over a period of time, focal adhesions develop. These are specializations of the plasma membrane where the cell is tightly apposed to the ECM, and are areas where integrin, talin, vinculin, and α-actinin colocalize (Burridge et al., 1988). Furthermore, the actin microfilaments are inserted at these focal adhesion sites, leading to the fully spread phenotype. The final phenotype is very variable depending on the cell type.

In migratory cells, these post-ligand events are probably similar to well spread nonmigratory cells, except that there must be a rapid on/off regulation of the integrin functions. Perhaps the best model of migration is that of the human neutrophils. These cells migrate very rapidly (10-20um/min; Maher et al., 1984) on many substrates, in both integrin-dependent and independent manners. When they are migrating on an integrin substrate, such as vitronectin, there are multiple intracellular fluxes of Ca^{2+} that correlate with the movements (Marks and Maxfield, 1990). Blockade of these transient fluxes with intracellular buffering, or by depletion of the extracellular Ca^{2+}, leads to a lack of movement (Marks et al., 1991). Furthermore, this appears to be due to the inability to detach from the substrate, since inclusion of peptides or antibodies to certain integrins lead to a restoration of movement (Hendey et al., 1992). In the case of VN, peptide inhibitors of calcineurin, a type 2b phosphatase, also blocks the movement (Hendey et al., 1992). These results are consistent with the conclusion that rises in intracellular Ca^{2+} allow activation of phosphatase 2b (which is calcium sensitive), followed by an unknown dephosphorylation event required for detachment. Further details of this mechanism await a better understanding of the molecular nature of the cytoskeletal apparatus interacting with integrins.

Recently, a number of approaches have been used to identify intracellular signals associated with integrin binding. For many years, there were several examples of differentiation events thought to be mediated by integrin binding to ligands. Among these, a particularly compelling example comes from studies of myocyte differentiation. *In vitro* myoblasts can be induced to fuse and differentiate when plated on integrin substrates and this differentiation is dependent on β_1 integrins (Menko and Boettinger, 1987).

Another example of an intracellular signal mediated by integrin have come from the studies of anchorage dependent growth, which many normal cells exhibit. Several labs have shown that these cells do not grow when placed in a suspension-like state (Benecke et al., 1978; DeLarco and Todaro, 1978). Recently, Guadagno and Assoian

(1991) have shown that this growth inhibition is due to a block in the cell cycle near the G1-S phase transition. This is near the START site in the yeast system. The start gene is a transcription factor necessary for transcription of genes require for DNA replication. Thus, it seems likely that integrin mediated attachment provides signals necessary for DNA replication in normal cells (such as activation of CDC2 kinase, or START functions). These signals are obviously bypassed in transformed cells, which can grow in suspension.

What is the nature of these signals and how are they generated? There are numerous signals generated via integrin attachment. All require clustering of integrins, either by antibody or by multivalent ligands (e.g., Schwartz et al., 1991b). This would occur in ECMs since most of the integrin substrates will be multivalent under those conditions. In several cell types, an activation of the Na/H antiport system occurs, which leads to a rise of intracellular pH (Schwartz et al., 1991a). This alkalinezation of the cytoplasm has been associated with growth, and indeed, there is an elevation of this pH in transformed fibroblasts. A second signal (as discussed previously) that can occur is increases in intracellular Ca^{2+} (Jaconi et al., 1991). This rise can be due to activation of calcium channels or to mobilization of intracellular stores.

Lastly, there has recently been a flurry of papers demonstrating tyrosine phosphorylation of specific cytoplasmic proteins upon integrins binding to ligands (Guan et al., 1991; Burridge et al., 1992) or clustering by antibodies (Kornberg et al., 1991). The major labeled protein is localized to focal adhesions in fibroblasts and has been termed focal adhesion kinase, or $pp125^{FAK}$ (Schaller et al., 1992). This intracellular kinase is phosphorylated rapidly upon integrin binding to ligand, and is a major phosphoprotein in most adherent cells. There are some indications that this protein might not be the most proximal signal in the integrin dependent pathway (Huang et al., 1993). Most of the data suggest that this phosphorylation is related to spreading, and while it may be required for proper adhesive functions, there may be other transient phosphoproteins involved in the initial stages of the pathway, proximal to the focal adhesion kinase. Again, in order to better understand this pathway, we need to know which proteins interact with the integrins within the cell. This brings us to a discussion of the cytoplasmic domains.

F. Cytoplasmic Domains

In general, the cytoplasmic domains of both the α and β subunits are quite short, from 13 (α_1) to 58 amino acids (β_7). The one exception is the β_4 cytoplasmic domain, which is larger than the extracellular portion of β subunits. The β_4 cytoplasmic domain is critical in the localization and function of this integrin in hemidesmosomes (Spinardi et al., 1993); however, I will not discuss it further. Overall, the β cytoplasmic domains are more highly conserved among types and between species than the α cytoplasmic domains. However, several α subunit cytoplasmic domains are virtually identical between species (i.e., α_1, α_3, α_5). When comparing among α subunits, there is a five amino acid stretch GFFKR (in single letter code) that is virtually identical in all α cytoplasmic domains. It is our experience (and others as well) that deletions or mutations of this domain lead to a lack of expression of the mutant α subunit, presumably due to an inability to dimerize with the corresponding β subunit. Thus, while this sequence is clearly required for expression of α subunits, its role in post-ligand effects remains unclear.

The nature of the molecules that interact with integrins on the cytoplasmic side of the cell remains unclear, despite intensive study. There are, however, certain points that can be made regarding the role of both α and β cytoplasmic domains in the functioning of integrins. After ligand has been bound by the receptor, a number of post-ligand events may occur. These include localization to focal adhesion sites, phosphorylation of putative signalling molecules, interaction with cytoskeletal components, and others as well. Most of the studies to date have focused on the role of integrin domains in the localization of these molecules to focal adhesion sites, where the plasma membrane of cells is in close apposition to the ECM. There is also colocalization at this point with the cytoskeleton. Initial studies showed that deletion of the β_1 cytoplasmic domain led to a loss of localization within the focal adhesion plaque (Solowska et al., 1989; Hayashi et al., 1990). Furthermore, Reszka and colleagues (1992) have shown that point mutations in three regions also affect the localization into focal adhesions. These studies all show that the β_1 cytoplasmic domain is required for focal contact localization. Furthermore, it has been shown recently that the β_1 cytoplasmic domain itself is able to localize a chimeric interleukin-β_1 protein to focal contacts. LaFlamme and colleagues (1992) constructed a chimeric protein consisting of the transmembrane and extracellular domain of the human interleukin-2 receptor and the integrin β_1 cytoplasmic domain. If cells expressing this protein are plated on FN, the interleukin-β_1 chimera colocalizes with endogenous FN receptors ($\alpha_5\beta_1$) in focal contacts. This finding suggests that this domain is not only necessary, but may be sufficient to localize integrin to focal contact sites. However, the behavior of integrin heterodimers *in vivo* is different than the chimera. A number of authors have shown that a specific integrin receptor does not localize to focal contacts unless it has bound ligand (Dejana et al., 1988; Singer et al., 1988). In addition, LaFlamme and colleagues (1992) showed that addition of a soluble ligand was effective in promoting receptor movement to focal contact sites, even when this ligand is not incorporated into the matrix. One explanation for these results would be that occupation of the integrin by ligand leads to a change in the receptor, by which the β cytoplasmic domain becomes "active" or "unmasked." Thus, we believe that there is an inhibitory conformation or component of the normal integrin receptor that prevents the β_1 cytoplasmic domain from being sufficient to localize to focal contact sites in the absence of ligand.

Recently, we have shown that the α cytoplasmic domain may play a role in the control of localization of integrin to focal contacts. Deletion of most of the α_1 cytoplasmic domain (preserving the GFFKR) leads to the localization of this integrin to focal contacts even on substrates which $\alpha_1\beta_1$ does not bind (Briesewitz et al., 1993b). Thus, loss of the α cytoplasmic domain leads to *ligand-independent* focal contact localization. These experiments suggest that in native integrins, the α cytoplasmic domain may "cover" the β_1 cytoplasmic domain, preventing ligand independent focal contact localization. Thus, we propose that in general, the α cytoplasmic domain has a modulating influence on β cytoplasmic domain function.

Recently, there have been a number of studies published which investigate the function of integrin α subunit cytoplasmic domains. Expression of the platelet fibrinogen receptor α_{IIb} β_3 in CHO cells leads to a latent receptor that can be activated with certain anti-receptor antibodies (O'Toole et al., 1990). This is similar to the situation in platelets, where activation via physiological mediators is required for activity. Expression of an α_{IIb} without its cytoplasmic domain in CHO cells leads to a constitutively active receptor (O'Toole et al., 1991), and substitution of the α_5 cytoplasmic domain for the native domain will not suppress this activity. However, these authors have interrupted the GFFKR sequence in these studies and more recently have shown that if this region is left intact, no activation occurs (Ylänne et al., 1993). In another system, Chan and colleagues (1992) have shown that the exchange of α cytoplasmic domain can change the functions of the respective integrins. These authors made chimeras of the extracellular portion of the integrin α_2 and various α subunit cytoplasmic domains. When the α_5 cytoplasmic domain was substituted for the α_2 cytoplasmic domain, collagen gel contraction via $\alpha_2\beta_1$ was unaffected. However, when the α_4 cytoplasmic domain was used, cells migrated on collagen but would not contract the gel. One explanation for these results might be that these different α cytoplasmic domains had differential effects on the "exposure" or "activation" state of the β_1 cytoplasmic domain.

Despite many differences in the amino acid sequences between various α cytoplasmic domains, a common mechanism of interaction between α and β cytoplasmic domains could exist. The recent finding of utilization of different cytoplasmic domains for the same α subunit via alternative splicing (Hogervorst et al., 1990; Tamura et al., 1991) suggests the possibility that modulation of β cytoplasmic domain function by α cytoplasmic domain is a critical point of integrin regulation.

The simplest mechanism in which α cytoplasmic domains could influence the β cytoplasmic domain would be if the two cytoplasmic domains would be near each other. Most models of integrin heterodimeric structures, however, show only the amino terminal segments of the extracellular domains of α and β subunits associated. As discussed previously, our data suggests that the cytoplasmic domains could be associated. More experiments are needed before we understand the mechanism of β cytoplasmic domain modulation.

In addition to mediating the many integrin post ligand binding events, the cytoplasmic domains of both the α and β subunits have been implicated in the inside-out signaling which occurs upon activation of integrins. As discussed earlier Hibbs and colleagues (1991) have shown that the c-terminal third of the the β_2 cytoplasmic domain contains three consecutive threonine residues, which are critical for activation. In addition, a recent report shows that a mutation in the β_3 cytoplasmic domain of a Glansmann thrombasthenia patient, where a proline is substituted for a serine leads to an inability to be activated by thrombin (Chen et al., 1992).

This serine is in a homologous position to the first of the consecutive threonines in the β_2 cytoplasmic domain. Furthermore, this region is very highly conserved among several integrin β subunits. These sequences would also be lost in the alternatively spliced variants of β_1 and β_3 (Balzac et al., 1993; van Kuppevelt et al., 1989).

As I stated earlier, it is unclear which molecules directly bind to integrins on the cytoplasmic aspect. There is *in vitro* evidence that both talin (Horwitz et al., 1986) and α-actinin (Otey et al., 1990) can bind to β cytoplasmic peptides. While both of these molecules are colocalized with integrins in focal adhesion sites, there is no *in vivo* evidence that they directly bind to integrins. In my opinion, the search is still on for molecules which interact *in vivo* with integrin cytoplasmic domains.

G. Final Model

I would like to propose a model of integrin function that is based on a large amount of data and a significant amount of speculation. A number of efforts to crystalize these molecules are underway, but most of them will be using soluble truncated (or proteolytically cleaved) versions of integrins. These versions will not fully recapitulate the nature of these receptors, due to the lack of the transmembrane regions that have been described previously. Furthermore, the apparently flexible nature of these receptors, which allow for regulation and multiple conformation states, may make organization into a crystal structure extremely difficult. In the interim, we will continue to propose models that will hopefully become more refined. Eventually, we all hope to have defined structural information.

This model is shown in Figure 2. In their resting state, most integrin receptors will have a relatively closed ligand binding pocket. In many cells, such as lymphocytes and platelets, this pocket will only be accessible to small peptides or extended portions of a fixed ligand. In state 2 the receptor could be viewed as having a partially active state. Now some ligands are accessible but others still are not. This is probably the state of the receptors on most adherent cells. The final state 3 is the fully activated receptor, which now can bind the full complement of ligands. Ligand binding in most cases leads to this state. These changes, we believe, induce a conformation change throughout the extracellular domains, which then affects the state

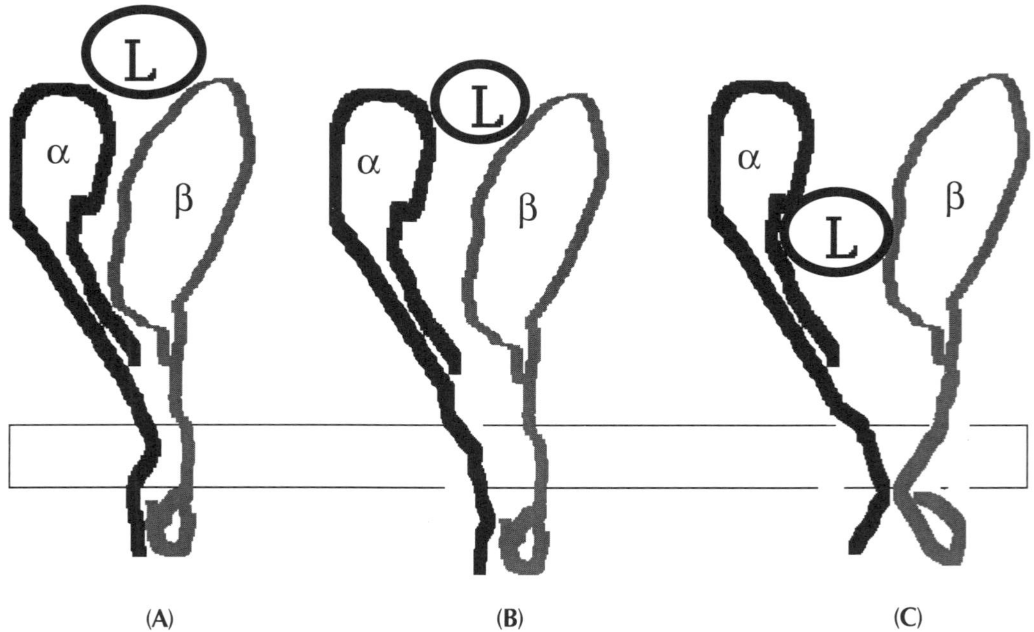

Figure 2. Model of integrin function. (**A**) Closed pocket which does not allow access to ligand (L). This is the situation in circulating (leukocytes, platelets) cells. An intracellular activation step is required for binding. The cytoplasmic domains are associated, blocking intracellular integrin function. (**B**) In adherent cells, this pocket may be more open. However, there may be an activation step required for high affinity binding. The cytoplasmic domains are still associated, making intracellular integrin function ligand-dependent. (**C**) Ligand has now bound and transduces a dissociation of α and β cytoplasmic domains, leading to intracellular integrin functions, such as signaling and cytoskeletal association. Inside-out signaling, causing extracellular activation could occur through the reverse of this process, with a dissociation of of cytoplasmic domains leading to a more open ligand binding pocket.

of the transmembrane domain. In this speculative model, the transmembrane and proximal cytoplasmic domains are viewed as a hinge through which conformation changes in the extracellular domain "activate" the β cytoplasmic domain, presumably by a loss of the normal inhibitory influence (the α cytoplasmic domain). In my view, activation, via inside-out signaling, could work also through this hinge, by affecting the relationship of the α and β cytoplasmic domains, and the transmembrane regions, which then leads to extracellular conformation changes. Fortunately, many of the elements of this model are testable and will no doubt be a major focus of research.

REFERENCES

Adams, J.C., & Watt, F. (1990). Changes in keratinocyte adhesion during terminal differentiation: Reduction in fibronectin binding precedes $\alpha 5\beta 1$ integrin loss from the cell surface. Cell 63, 425-435.

Alcover, A., Mariuzza, R.A., Ermonval, M., & Acuto, O. (1990). Lysine 271 in the transmembrane domain of the T-cell antigen receptor β chain is necessary for its assembly with the CD3 complex but not for α/β dimerization. J. Biol. Chem. 265, 4131-4135.

Altruda, F., Cervella, P., Tarone, G., Botta, C., Balzac, F., Sefanuto, G., & Silengo, L. (1990). A human integrin β_1, subunit with a unique cytoplasmic domain generated by alternative mRNA processing. Gene 95, 261-266.

Argraves, W.S., Suzuki, S., Arai, H., Thompson, K., Pierschbacher, M.D., & Ruoslahti, E. (1987). Amino acid sequence of the human fibronectin receptor. J. Cell Biol. 105, 1183-1190.

Balzac, F., Belkin, A.M., Koteliansky, V.E., Blabanov, Y.V., Altruda, F., Silengo, L., & Tarone, G. (1993). Expression and functional analysis of a cytoplasmic domain variant of the β_1, integrin subunit. J. Cell. Biol. 121, 171-178.

Benecke, B.J., Ben-Ze'ev, A., & Penman, S. (1978). The control of mRNA production, translation, and turnover in suspended and reattached anchorage-dependent fibroblasts. Cell 14, 931-939.

Bennett, J. S., Kolodziej, M. A., Vilaire, G., & Poncz, M. (1993). Determinants of the intracellular fate of truncated forms of the platelet glycoproteins IIb and IIIa. J. Biol. Chem. 268, 3580-3585.

Blumberg, R.S., Alarcon, B., Sancho, J., McDermott, F.V., Lopez, P., Breitmeyer, J., & Terhorst, C. (1990). Assembly and function of the T cell antigen receptor. Requirement of either the lysine or arginine residues in the transmembrane region of the α chain. J. Biol. Chem. 265, 14036-14043.

Bonifacino, J.S., Cosson, P., & Klausner, R. D. (1991a). Colocalized transmembrane determinants for ER degradation and subunit assembly explain the intracellular fate of TCR chains. Cell 63, 503-513.

Bonifacino, J.S., Cosson, P., Shah, N., & Klausner, R.D. (1991b). Role of potentially charged transmembrane residues in targeting proteins for retention and degradation within the endoplasmic reticulum. EMBO J. 10, 2783-2793.

Brass, L.F., & Shattil, S.J. (1984). Identification and function of the high affinity binding sites for Ca^{2+} on the surface of platelets. J. Clin. Invest. 73, 626-632.

Briesewitz, R., Epstein, M.R., & Marcantonio, E.E. (1993a). The expression of native and truncated forms of the human integrin α_1 subunit. J. Biol Chem. 268, 2989-2996.

Briesewitz, R., Kern, A., & Marcantonio, E. E. (1993b). Ligand-dependent and independent integrin focal contact localization: The role of the α chain cytoplasmic domain. Mol. Biol. Cell. 4, 593-604.

Briesewitz, R., Kern, A., & Marcantonio, E.E. (1995). Assembly and function of integrin receptors is dependent on opposing α and β cytoplasmic domains. Mol. Biol. Cell 6, 997-1010.

Burridge, K., Fath, K., Kelly, T., Nuckolls, G., & Turner, C. (1988). Focal adhesion: Transmembrane junctions between the extracellular matrix and the cytoskeleton. Annu. Rev. Cell Biol. 4, 487-525.

Burridge, K., Turner, C.E., & Romer, L.H. (1992). Tyrosine phosphorylation of paxillin and $pp125^{FAK}$ accompanies cell adhesion to extracellular matrix: A role in cytoskeletal assembly. J. Cell. Biol. 119, 893-903.

Busk, M., Pytela, R., & Sheppard, D. (1992). Characterization of the integrin $\alpha_v\beta_6$ as a fibronectin-binding protein. J. Biol. Chem. 267, 5790-5796.

Buyon, J. P., Slade, S. G., Reibman, J., Abramson, S. B., Philips, M. R., Weissman, G., & Wincester, R. (1990). Constitutive and induced phosphorylation of the α and β chains of the CD11/CD18 leukocyte integrin family. Relationship to adhesion-dependent functions. J. Immunol. 144, 191-197.

Chan, B.M.C., & Hemler, M.E. (1993). Multiple functional forms of the integrin VLA-2 can be derived from a single $\alpha2$ cDNA clone: Interconversion of forms induced by an anti-$\beta1$ antibody. J. Cell. Biol. 120, 537-543.

Chan, B. M. C., Kassner, P. D., Schiro, J. A., Byers, H. R., Kupper, T. S., & Hemler, M. E. (1992). Distinct cellular functions mediated by different VLA integrin α subunit cytoplasmic domains. Cell 68, 1051-1060.

Chatila, T.A., Geha, R.S., & Arnaout, M.A. (1989). Constitutive and stimulus-induced phosphorylation of CD11/CD18 leukocyte adhesion molecules. J. Cell. Biol. 109, 3435-3444.

Chen, Y.P., Djaffer, I., Pidard, D., Steiner, B., Cieutat, A.M., Caen, J.P., & Rosa, J.P. (1992). Ser-752-Pro mutation in the cytoplasmic domain of integrin $\beta3$ subunit and defective activation of platelet integrin $\alpha IIb\beta3$ (glycoprotein IIb-IIIa) in a variant of Glanzmann thrombasthenia. Proc. Natl. Acad. Sci. USA 89, 10169-10173.

Cheresh, D.A., & Spiro, R.C. (1987). Biosynthetic and functional properties of an Arg-Gly-Asp-directed receptor involved in human melanoma cell attachment to vitronectin, fibrinogen, and von Willebrand factor. J. Biol. Chem. 262, 17703-17711.

Cheresh, D.A., Pytela, R., Pierschbacher, M.D., Klier, F.G., Ruoslahti, E., & Reisfeld, R.A. (1987). An arg-gly-asp-directed receptor on the surface of human melanoma cells exists in a divalent cation-dependent functional complex with the disialoganglioside GD2. J. Cell. Biol. 105, 1163-1173.

Conforti, G., Zanetti, A., Pasquali-Ronchetti, I., Quaglino, D., Neyroz, P., & Dejana, E. (1990). Modulation of the vitronectin receptor binding by membrane lipid composition. J. Biol. Chem. 265, 4011-4019.

Corbi, A.L., Kishimoto, T.K., Miller, L.J., & Sprinter, T.A. (1988). The human leukocyte adhesion glycoprotein Mac-1 (Complement receptor type 3, CD11b), α subunit. Cloning, primary structure, and relation to the integrins, von Willebrand factor and factor B. J. Biol. Chem. 263, 12403-12411.

D'Souza, S.E., Ginsberg, M.H., Burke, T.A., Lam, S.C.-T., & Plow, E.F. (1988). Localization of an Arg-Gly-Asp recognition site within an integrin adhesion receptor. Science 242, 91-93.

D'Souza, S.E., Ginsberg, M. H., Matsueda, G.R., & Plow, E. F. (1991). A discrete sequence in a platelet integrin is involved in ligand recognition. Nature 350, 66-68.

Dana, N., Fathallah, D.M., & Arnaout, M.A. (1991). Expression of a soluble and functional form of the human beta 2 integrin CD11b/CD18. Proc. Natl. Acad. Sci. USA 88, 3106-3110.

Dejana, E., Colella, S., Conforti, G., Abbadini, M., Gaboli, M., & Marchisio, P.C. (1988). Fibronectin and vitronectin regulate the organization of their respective Arg-Gly-Asp adhesion receptors in cultured human endothelial cells. J. Cell Biol. 107, 1215-1223.

DeLarco, J.E., & Todaro, G. (1978). Growth factors from murine sarcoma virus-transformed cells. Proc. Natl. Acad. Sci. USA 75, 4001-4005.

DeSimone, D.W., & Hynes, R.O. (1988). *Xenopus laevis* integrins: Structural conservation and evolutionary divergence of integrin β subunits. J. Biol. Chem. 263, 5333-5340.

Diamond, M.S., Garcia-Aguilar, J., Bickford, J.K., Corbi, A.L., & Springer, T.S. (1993). The I-domain is a major recognition site on the leukocyte integrin Mac-1 (CD11b/CD18) for four distinct adhesion ligands. J. Cell Biol. 120, 1031-1043.

Dransfield, I., Cabanas, C., Craig, A., & Hogg, N. (1992). Divalent cation regulation of the function of the leukocyte integrin LFA-1. J. Cell Biol. 116, 219-226.

Duperray, A., Troesch, A., Berthier, R., Chagnon, E., Frachet, F., Uzan, G., & Marguerie, G. (1989). Biosynthesis and assembly of platelet GPIIb-IIIa in human megakaryocytes: Evidence that assembly between pro-GPIIb and GPIIIa is a prerequisite for expression of the complex on the cell surface. Blood 74, 1603-1611.

Elices, M.J., & Hemler, M.E. (1989). The human integrin VLA-2 is a collagen receptor on some cells and a collagen/laminin receptor on others. Proc. Natl. Acad. Sci. USA 86, 9906-9910.

Frachet, P., Duperray, A., Delachanal, E., & Marguerie, G. (1992). Role of the transmembrane and cytoplasmic domains in the assembly and surface exposure of the platelet integrin GPIIb/IIIa. Biochemistry 31, 2408-2415.

Gailit, J., & Ruoslahti, E. (1988). Regulation of the fibronectin receptor affinity by divalent cations. J. Biol. Chem. 263, 12927-12932.

Geiger, B., Salomon, D., Takeichi, M., & Hynes, R. O. (1992). A chimeric N-cadherin/β_1-integrin receptor which localizes to both cell-cell and cell-matrix adhesions. J. Cell. Sci. 103, 943-951.

Grzesiak, J.J., Davis, G.E., Kirchhofer, D., & Pierschbacher. (1992). Regulation of $\alpha_2\beta_1$-mediated fibroblast migration on type I collagen by shifts in the concentrations of extracellular Mg^{2+} and Ca^{2+}. J. Cell Biol. 117, 1109-1118.

Guadagno, T.M., & Assoian, R.K. (1991). G1/S control of anchorage-independent growth in the fibroblast cell cycle. J. Cell. Biol. 115, 1419-1425.

Guan, J., Trevithick, J.E., & Hynes, R.O. (1991). Fibronectin/integrin interaction induces tyrosine phosphorylation of a 120 kDa protein. Cell Reg. 2, 951-964.

Hayashi, Y., Haimovich, B., Rezka, A., Boettiger, D., & Horwitz, A. F. (1990). Expression and function of chicken integrin $\beta 1$ subunit and its cytoplasmic domain mutants in mouse NIH 3T3 cells. J. Cell Biol. 110, 175-184.

Hemler, M.E. (1990). VLA proteins in the integrin family: Structures, functions, and their role on leukocytes. Annu. Rev. Immunol 8, 365-400.

Hendey, B., Klee, C. B., & Maxfield, F.R. (1992). Inhibition of neutrophil chemokinesis on vitronectin by inhibitors of calcineurin. Science 258, 296-299.

Hermanowski-Vosatka, A., Van Strijp, J.A.G., Swiggard, W.J., & Wright, S.D. (1992). Integrin modulating factor-1: A lipid that alters the function of leukocyte integrins. Cell 68, 341-352.

Hibbs, M.L., Jakes, S., Stacker, S.A., Wallace, R.W., & Springer, T.A. (1991). The cytoplasmic domain of the integrin lymphocyte function-associated antigen 1β subunit: Sites required for binding to inter-cellular adhesion molecule 1 and the phorbol ester-stimulated phosphorylation site. J. Exp. Med. 174, 1277-1238.

Hillery, C.A., Smyth, S.S., & Parise, L.V. (1991). Phosphorylation of human platelet glycoprotein llla(GPllla). Dissociation from fibrinogen receptor activation and phosphorylation of GPllla *in vitro*. J. Biol. Chem. 266, 14663-14669.

Hirst, R., Horwitz, A., Buck, C., & Rohrschneider, L. (1986). Phosphorylation of the fibronectin receptor complex in cells transformed by oncogenes that encode tyrosine kinases. Proc. Natl. Acad. Sci. USA 83, 6470-6474.

Ho, M.K., & Springer, T.A. (1983). Biosynthesis and assembly of the α and β subunits of Mac-1, a macrophage glycoprotein associated with complement receptor function. J. Biol Chem. 258, 2766-2769.

Hogervorst, F., Kuikman, I., von den Borne, A. E. G. K., & Sonnenberg, A. (1990). Molecular cloning of the human α_6 integrin subunit. Alternative splicing of the α_6 mRNA and chromosomal localization of the α_6 and β_4 genes. Eur. J. Biochem. 199, 425-433.

Horvath, A.R., Elmore, M.A., & Kellie, S. (1990). Differential tyrosine-specific phosphorylation of integrin in Rous sarcoma virus transformed cells with differing transformed phenotypes. Oncogene 5, 1349-1357.

Horwitz, A. F., Duggan, E., Buck, C., Beckerle, M. C., & Burridge, K. (1986). Interaction of plasma membrane fibronectin receptor with talin—a transmembrane linkage. Nature 320, 531-533.

Huang, M.M., Lipfest, L., Cunningham, M., Brugge, J.S., Ginsberg, M.H., & Shattil, S.J. (1993). Adhesive ligand binding to Integrin $\alpha IIb\beta_3$ stimulates tyrosine phosphorylation of pp 125^{FAK}. J. Cell Biol. 122, 473-483.

Hynes, R.O. (1992). Integrins: Versatility, modulation and signalling in cell adhesion. Cell 69, 11-25.

Hynes, R.O. (1987). Integrins: A family of cell surface receptors. Cell 48, 549-554.
Jaconi, M.E.E., Theler, J.M., Schlegel, W., Appel, R.D., Wright, S.D., & Lew P.D. (1991). Multiple elevations of cytosolic-free Ca^{2+} in human neutrophils: Initiation by adherence receptors of the integrin family. J. Cell. Biol. 112, 1249-1257.
Kirchhofer, D., Languino, L.R., Ruoslahti, E., & Pierschbacher, M.D. (1990). $\alpha_2\beta_1$ integrins from different cell types show different binding specificities. J. Biol. Chem. 265, 615-618.
Kishomoto, T.K., O'Connor, K., & Springer, T.A. (1989). Leukocyte adhesion deficiency. Aberrant splicing of a conserved integrin sequence causes a moderate deficiency phenotype. J. Biol. Chem. 264, 3588-3595.
Kornberg, L.J., Earp, H.S., Turner, C.T., Prockop, C., & Juliano, R.L. (1991). Signal transduction by integrins: Increased protein tyrosine phosphorylation caused by clustering of $\beta 1$ integrins. PNAS 88, 8392-8396.
Kretsinger, R.H. (1980). Structure and evolution of calcium-modulated proteins. CRC Crit. Rev. Biochem. 8, 119-174.
LaFlamme, S.E., Akiyama, S.K., & Yamada, K.M. (1992). Regulation of fibronectin receptor distribution. J. Cell Biol. 117, 437-447.
Lanza, F., Kieffer, N., Phillips, D. R., & Fitzgerald, L.A. (1990). Characterization of the human platelet glycoprotein IIIa gene. Comparison with the fibronectin receptor beta-subunit gene. J. Biol. Chem. 265, 18098-18103.
Loftus, J.C., O'Toole, T.E., Plow, E.F., Glass, A., Frelinger, A.L., & Ginsberg, M.H. (1990). A beta 3 integrin mutation abolishes ligand binding and alters divalent cation-dependent conformation. Science 249, 915-918.
Maher, J., Martell, J.V., Brantley, B.A., Cox, E.B., Niedel, J.E., & Rosse, W.F. (1984). The response of human neutrophils to a chemotactic tripeptide (N-formyl-methionyl-leucyl-phenylalanine) studied by microcinematography. Blood 64, 221-228.
Marcantonio, E.E., & Hynes, R.O. (1988). Antibodies to the conserved cytoplasmic domain of the integrin β_1 subunit react with proteins in vertebrates, invertebrates and fungi. J. Cell Biol. 106, 1765-1772.
Marcantonio, E.E., Guan, J., Trevethick, J.E., & Hynes, R.O. (1990). Mapping of the functional determinants of the integrin β_1 cytoplasmic domain by site directed mutagenesis. Cell Regulation 1, 597-604.
Marks, P. W., & Maxfield, F. R. (1990). Transient increases in cytosolic free calcium appear to be required for the migration of adherent human neutrophils. J. Cell. Biol. 110, 43-52.
Marks, P. W., Hendey, B., & Maxfield, F. R. (1991). Attachment to fibronectin or vitronectin makes human neutrophil migration sensitive to alterations in cytosolic free calcium concentration. J. Cell. Biol. 112, 149-158.
Menko, A.S., & Boettinger, D. (1987). Occupation of the extracellular matrix receptor, integrin, is a control point for myogenic differentiation. Cell 51, 51-57.
Michishita, M., Videm, V., & Arnaout, M.A. (1993). A novel divalent cation binding site in the A domain of the $\beta 2$ integrin CR3 (CD11b/CD18) is essential for ligand binding. Cell 72, 857-867.

Neugebauer, K.M., & Reichardt, L.F. (1991). Cell-surface regulation of β1-integrin activity on developing retinal neurons. Nature 350, 68-71.

O'Toole, T.E., Loftus, J. C., Du, X. P., Glass, A. A., Ruggeri, Z. M., Shattil, S. J., Plow, E. F., & Ginsberg, M. H. (1990). Affinity modulation of the $\alpha_{IIb}\beta_3$ integrin is an intrinsic property of the receptor. Cell Regul. 1, 883-893.

O'Toole, T.E., Mandelman, D., Forsyth, J., Shattil, S. J., Plow, E. F., & Ginsberg, M. H. (1991). Modulation of the affinity of integrin $\alpha_{IIb}\beta_3$ (GP IIb-IIIa) by the cytoplasmic domain of α_{IIb}. Science 254, 845-847.

Otey, C.A., Pavalko, F. M., & Burridge, K. (1990). An interaction between α-actinin and the β_1 integrin subunit *in vitro*. J. Cell Biol. 111, 721-729.

Poncz, M., Eisman, R., Heidenreich, R., Silver, S.M., Vilaire, G., Surrey, S., Schwartz, E., & Bennett, J.S. (1987). Structure of the platelet membrane glycoprotein IIb. Homology to the alpha subunits of the vitronectin and fibronectin membrane receptors. J. Biol. Chem. 262, 8476-8482.

Pullman, W.E., & Bodmer, W.F. (1992). Cloning and characterization of a gene that regulates cell adhesion. Nature 356, 529-532.

Pytela, R. (1988). Amino acid sequence of the murine Mac-1 α chain reveals homology with the integrin family and an additional domain related to von Willebrand factor. E.M.B.O. J. 7, 1371-1378.

Reszka, A. A., Hayashi, Y., & Horwitz, A.F. (1992). Identification of amino acid sequences in the integrin β_1 cytoplasmic domain implicated in cytoskeletal association. J. Cell. Biol. 117, 1321-1329.

Rosa, J.P., & McEver, R.P. (1989). Processing and assembly of the integrin, glycoprotein IIb-IIIa, in HEL cells. J. Biol. Chem. 264, 12596-12603.

Rothleim, R., & Springer, T.A. (1986). The requirement for lymphocyte function-associated antigen 1 in homotypic leukocyte adhesion stimulated by phorbol ester. J. Exp. Med. 163, 1132-1149.

Schaller, M.D., Borgman, C.A., Cobb, B.S., Vines, R.R., Reynolds, A.B., & Parsons, J.T. (1992). $PP125^{FAK}$, a structurally unique protein tyrosine kinase associated with focal adhesions. Proc. Natl. Acad. Sci. USA 89, 5192-5196.

Schwartz, M.A., Ingber, D.E., Lawrence, M., Springer, T.A., & Lechene, C. (1991). Multiple integrins share the ability to induce elevation of intracellular pH. Exp. Cell. Res. 195, 533-535.

Schwartz, M.A., Lechene, C., & Ingber, D.E. (1991b). Insoluble fibronectin activates the Na/H antiporter by clustering and immobilizing integrin $\alpha_5\beta_1$, independent of cell shape. Proc. Natl. Acad. Sci. USA 88, 7849-7853.

Shih, D.-T., Edelman, J.M., Horwitz. A.F., Grunwald, G.B., & Buck, C.A. (1993). Structure/function analysis of the integrin β_1 subunit by epitope mapping. J. Cell Biol. 122, 1361-1371.

Singer, I. I., Scott, S., Kawka, D. W., Kazazis, D. M., Gailit, J., & Rouslahti, E. (1988). Cell surface distribution of fibronectin and vitronectin receptor depends on substrate composition and extracellular matrix accumulation. J. Cell. Biol. 106, 2172-2182.

Smith, J.W., & Cheresh, D.A. (1988). The Arg-Gly-Asp binding domain of the vitronectin receptor. Photoaffinity cross-linking implicates amino acid residues 61-203 of the β subunit. J. Biol. Chem. 263, 18726-18731.

Smith, J.W., & Cheresh, D.A. (1990). Integrin ($\alpha_v\beta_3$)-ligand interaction. J. Biol. Chem. 265, 2168-2172.

Smyth, S. S., Hillery, C. A., & Parise, L. V. (1992). Fibrinogen binding to purified platelet glycoprotein IIb-IIIa (integrin alpha IIb beta 3) is modulated by lipids. J. Biol. Chem. 267, 15568-15577.

Solowska, J., Guan, J-L., Marcantonio, E.E., Trevithick, J.E., Buck, C.A., & Hynes, R.O. (1989). Expression of normal and mutant avian integrin subunits in rodent cells. J. Cell. Biol. 109, 853-861.

Spinardi, L., Ren, Y.L., Sanders, R., & Giancotti, F.G. (1993). The β_4 subunit cytoplasmic domain mediates the interaction of $\alpha_6\beta_4$ integrin with the cytoskeleton of hemidesmosomes. Mol. Biol. Cell. 4, 871-884.

Springer, T.A. (1990). Adhesion receptors of the immune system. Nature 346, 425-434.

Staatz, W.D., Rajpara, S.R., Wayner, E.A., Carter, W.C., & Santoro, S.A. (1989). The membrane glycoprotein Ia-IIa (VLA-2) complex mediates the Mg^{2+}-dependent adhesion of platelets to collagen. J. Cell. Biol. 108, 1917-1924.

Takada, Y., & Hemler, M.E. (1989). The primary structure of the VLA-2/collagen receptor α^2 subunit (platelet GP Ia): Homology to other integrins and the presence of a possible collagen-binding domain. J. Cell Biol. 109, 397-407.

Takada, Y., Ylänne, J., Mandelman, D., Puzon, W., & Ginsberg, M. H. (1992). A point mutation of integrin β_1 subunit blocks binding of $\alpha_5\beta_1$ to fibronectin and invasin but not recruitment to adhesion plaques. J. Cell Biol. 119, 913-921.

Tamkun, J.W., DeSimone, D.W., Fonda, D., Patel, R.S., Buck, C., Horwitz, A.F., & Hynes, R.O. (1986). Structure of integrin, a glycoprotein involved in the transmembrane linkage between fibronectin and actin. Cell 42, 271-282.

Tamura, R.N., Rozzo, C., Starr, L., Chambers, J., Reichardt, L. F., Cooper, H. M., & Quarenta, V. (1991). Cell-type specific integrin variants with alternative α chain cytoplasmic domains. Proc. Natl. Acad. Sci. USA 88, 10183-10187.

van Kuppevelt, T.H.M.S.M., Languino, L.R., Gailit, J.O., Suzuki, S., & Ruoslahti, E. (1989). An alternative cytoplasmic domain of the integrin β_3 subunit. Proc. Natl. Acad. Sci. USA 86, 5415-5418.

Vogel, B.E., Tarone, G., Giancotti, F.G., Gailit, J., & Ruoslahti, E. (1990). A novel fibronectin receptor with an unexpected subunit composition ($\alpha_v\beta_1$). J. Biol. Chem. 265, 5934-5937.

Ylänne, J., Chen, Y.L., O'Toole, T.E., Loftus, J.C., Takada, Y., & Ginsberg, M.H. (1993). Distinct functions of integrin α and β subunit cytoplasmic domains in cell spreading and formation of focal adhesions. J. Cell. Biol. 122, 223-234.

Zusman, S., Grinblat, Y., Yee, G., Kafatos, F.C., & Hynes, R.O. (1993). Analyses of PS Integrin functions during *Drosophila* development. Development 118, 737-750.

FUNCTION AND REGULATION OF SELECTINS:

A NEW FAMILY OF LEUKOCYTE AND ENDOTHELIAL CELL ADHESION PROTEINS

Mark A. Jutila

Advances in Molecular and Cell Biology,
Volume 16, pages 31-61.

ISBN: 0-7623-0143-0.

I. INTRODUCTION

Adhesive interactions between leukocytes and endothelial cells, leukocytes and other leukocytes, and leukocytes and platelets are required and occur in all types of inflammation. A new family of adhesion proteins called Selectins is expressed by all three cell types and is thought to be critically important in regulation adhesive interactions between them (Figure 1). Three members of this family have been defined: One is expressed by leukocytes (L-selectin), another by endothelial cells (E-selectin), and the third by both endothelial cells and platelets (P-selectin). The relationship of these molecules only became evident after the cloning of their cDNAs. Each molecule consists of a N-terminal domain homologous to C-type lectins, a short sequence similar to epidermal growth factor (EGF), multiple short consensus repeats (SCRs) similar to complement binding-like proteins, a transmembrane domain, and a short cytoplasmic tail. Overall, the selectins exhibit 40-60 percent identity at the amino acid level. The main difference in the proteins is in the number of SCRs; L-, E-, and P-selectin have two, six, and nine, respectively. The identification of a common lectin domain suggests that carbohydrate recognition, which was previously shown to be important only for L-selectin, is a feature of all three molecules.

Models based on *in vivo* and *in vitro* studies have been proposed for the function of selectins during inflammation. Intravital microscopy has shown that leukocyte extravasation can be separated into three distinct steps: (1) a reversible rolling interaction between the leukocyte and the vascular endothelium, (2) tight adhesion, and (3) transendothelial cell migration. All three steps are essential for effective leukocyte accumulation at sites of inflammation. The vascular selectins as well as L-selectin are thought to function predominantly during the first interaction, whereas integrins (CD11a-c/CD18, VLA4) and their counter-receptors (ICAM-1, ICAM-2, ICAM-3, and VCAM-1) control tight adhesion and transendothelial cell migration. The VLA4/VCAM-1 interaction

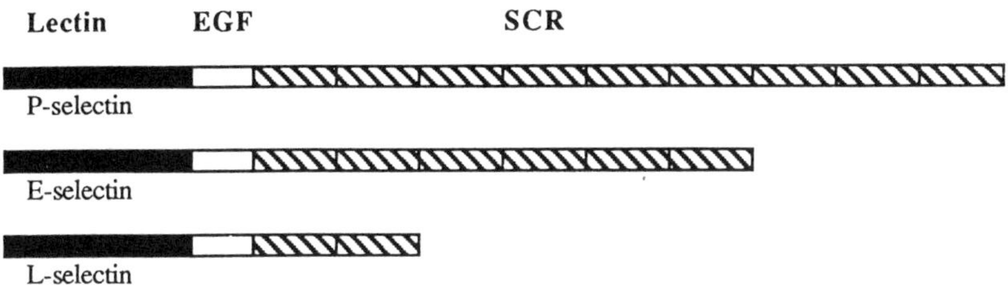

Notes: P-selectin: Previously called PADGEM or GMP-140. Expressed by thrombin activated platelets and endothelial cells. Supports binding of myeloid cells via recognition of SLe^x on the leukocyte surface.

E-selectin: Previously called ELAM-1. Expressed within two to four hrs after cytokine stimulation of endothelial cells. Supports binding of a myeloid cells and subsets of lymphocytes. Like P-selectin, E-selectin recognizes SLe^x, plus other structurally related carbohydrates.

L-selectin: Previously called peripheral-lymph-node homing receptor, gp90MEL-14, LECAM-1, and LAM-1. Expressed by all circulating leukocytes, except certain subsets of memory lymphocytes. Regulates lymphocyte recirculation and leukocyte binding to cytokine activated endothelial cells. Recognizes sialic acid modified carbohydrates on endothelial cells.

Key:
- Domain homologous to mammalian C-type lectins.
- Domain homologous to epidermal growth factor (EGF).
- Short consensus repeats (SCRs) homologous to complement binding proteins.

Figure 1. Selectins

may also contribute to rolling, but this interaction has not been fully characterized (Wolber et al., 1993). The transition of a selectin-dependent rolling interaction to the integrin-dependent event requires activating signals, which are likely delivered by chemotactic factors released by the inflamed tissues. A more complete description of this model is presented by Butcher (1991).

The intent of this chapter is to provide a brief historical review of the selectins, followed by a discussion of some potential future research directions on these molecules. Advances up through 1993 will be emphasized and speculation on novel selectin receptors will be presented. This chapter is not intended as a comprehensive review of the selectin literature and apologies are extended to those whose excellent contributions on selectins are not covered. Furthermore, because of limits in modifications at the proof stage, only a few of the many recent advances are discussed.

II. SELECTINS

A. L-selectin

L-selectin (previously called peripheral-lymph-node homing receptor, gp90MEL-14, LECAM-1, LAM-1), originally characterized as a 90 kD lymphocyte glycoprotein, specifically regulates lymphocyte recirculation through peripheral lymphoid tissues by binding to specialized, postcapillary venules called high-walled endothelial venules (HEV). L-selectin was shown to be minimally involved in the recirculation of lymphocytes through other lymphoid tissues, such as the gut; thus, it was the first example of a tissue-specific adhesion receptor. (The reader is referred to selected articles that more adequately review lymphocyte homing receptors, which is beyond the intent of this chapter, e.g., Berg et al., 1989; Butcher, 1986; Picker and Butcher, 1992; Rosen, 1989). L-selectin was first defined in the mouse, but has now been characterized in humans, cows, sheep, pigs, goats, rats, and dogs (Kishimoto et al., 1990a; Tedder et al., 1989; Spertini et al., 1991b; Walcheck et al., 1992a; Jutila et al., 1992; Abbassi et al., 1991). Analysis at the functional, biochemical, and molecular level in all these animals shows that L-selectin is an evolutionarily well-conserved molecule, suggesting that it must be important for survival.

L-selectin is not only expressed by lymphocytes, but is also found on neutrophils, monocytes, and eosinophils (Lewinsohn et al., 1987). The molecule exhibits different molecular weights on these other leukocytes, which is due to differences in glycosylation and not in primary amino acid sequence (Ord et al., 1990). L-selectin on monoyctes and neutrophils can bind peripheral lymph node HEV, but the primary function of the myeloid molecule appears to be during the interaction of these cells with inflamed endothelium. Anti-L-selectin antibodies block neutrophil adhesion to cytokine-stimulated endothelial cells in assays done under shear or flow (Hallmann et al., 1991; Kishimoto et al., 1990b; Spertini et al., 1991a; Smith et al., 1991; Abbassi et al., 1993; Bargatze et al., 1994).

The function of L-selectin has also been examined *in vivo*. Anti-L-selectin mAb specifically block lymphocyte recirculation through peripheral lymph nodes in mice (Gallatin et al., 1983). Further, anti-L-selectin antibodies and soluble L-selectin itself block neutrophil (Lewinsohn et al., 1987; Jutila et al., 1989; Watson et al., 1991a) and monocyte (Jutila et al., 1990) accumulation at sites of acute inflammation in the skin and peritoneal cavity of mice. A number of abstracts have been presented at recent meetings which show that inhibitors of L-selectin are effective at blocking diverse inflammatory events in the lungs (Mulligan et al., 1993). L-selectin chimeras also block inflammatory cell recruitment (Mulligan et al., 1993). Recent gene knock-out studies have confirmed the importance of L-selectin in regulating lymphocyte homing to peripheral lymph node sites of inflammation (Arbones et al., 1994). Thus, L-selectin has been shown to mediate diverse endothelial cell/leukocyte interactions *in vitro* and inhibitors of L-selectin block leukocyte migration into lymphoid tissues and sites of inflammation *in vivo*.

B. E-selectin

E-selectin, originally called endothelial-cell leukocyte adhesion molecule-1 (ELAM-1), was first described as a 110 kD glycoprotein expressed on cytokine-activated, cultured endothelial cells and on venules in sites of inflammation in vivo. Endothelial cells that express E-selectin support the binding of myeloid cells, such as neutrophils, and the binding can be blocked by anti-E-selectin mAbs (Bevilacqua et al., 1987). Recently, a subset of human lymphocytes was shown to bind E-selectin (Graber et al., 1990; Picker et al., 1991b; Shimizu et al., 1991; Postigo et al., 1992). These lymphocytes represent memory cells, which are specifically recognized by the HECA 452 mAb (Shimizu et al., 1991; Picker et al., 1991b). The expression of the HECA 452 antigen (cutaneous lymphocyte-associated antigen, CLA), following activation induced by antigen stimulation, correlates precisely with the capacity to bind E-selectin and may serve as an E-selectin binding epitope (see following; Picker et al., 1993b).

While most conventional T cells require an activating and/or differentiation signal to acquire the capacity to bind E-selectin, we have found that certain specialized subsets do not. γ/δ T cells, which are distinguished from other T cells (α/β T cells) by the genes that encode their surface receptor for antigen (reviewed in Allison and Havran, 1991), exhibit homing patterns similar to those of memory T cells (Mackay, 1991). However, unlike memory T cells, this homing phenotype is seen with cells from newborn animals. In a recent report, we have shown that γ/δ T cells from newborn calves avidly bind E-selectin in in vitro binding assays (Walcheck et al., 1993), suggesting that differentiation into memory cells is not required for their binding or the memory phenotype of bovine γ/δ T cells is attained before birth. Interestingly, the HECA 452 mAb does not recognize these T cells, showing that even though the expression of this antigenic epitope correlates with binding in the human, it is not absolutely required for lymphocyte E-selectin interactions.

In vivo, E-selectin expression is associated with many types of inflammatory reactions and the accumulation of specific leukocyte subsets which have been shown to bind E-selectin in *in vitro* assays. E-selectin was first demonstrated *in vivo* on venules at a site of delayed-type hypersensitivity reaction in humans. Though usually restricted to postcapillary, it has also been found on capillary endothelial cells in certain animal models (Cotran, 1987; Pober, 1988; Pober and Cotran, 1991; Leung et al., 1991; Munro et al., 1991). Kinetic analysis shows that induction of E-selectin correlates with an influx of neutrophils into dermal sites of inflammation (Munro et al., 1991). Picker and colleagues (1991b) found that even though E-selectin can be found in virtually any type of acute inflammatory lesion, in chronic inflammation it exhibits a biased expression in the skin. The expression of E-selectin in the skin correlates with the presence of large numbers of CLA-positive lymphocytes.

The biased chronic expression of E-selectin in the dermis suggests that it may be important for the trafficking of skin-seeking lymphocytes (Picker et al., 1991b; MacKay, 1991). We have recently tested whether this hypothesis is consistent with the in vivo migration of γ/δ T cells, which avidly bind E-selectin. We found that injection of 1 ug of TNF into the skin of calves dramatically increases expression of E-selectin on venules at the injection site which correlates with almost a 10-fold increase in the numbers of γ/δ T cells within the tissue (Walcheck et al., 1993). In subsequent studies, we have found that by inducing a DTH reactivity to purified protein derivative or by injecting low doses of LPS we can induce a similar correlation between increased E-selectin expression and the recruitment of γ/δ T cells (Jutila, unpublished).

The evaluation of the effect of E-selectin inhibitors *in vivo* has been slow, due in part to the lack of reagents in appropriate animal models. Recently, cDNAs for both vascular selectins were cloned in the mouse (Sanders et al., 1992; Weller et al., 1992), thus, in the near future, effective anti-selectin reagents may be available for studies in mice. Some reagents do crossreact in rats and Mulligan and colleagues (1991) have shown that anti-E-selectin mAbs block neutrophil accumulation into inflamed peritoneum and lungs. Gundel and colleagues (1991) have used anti-human, E-selectin mAbs that crossreact in the monkey to block the late-phase airway obstruction during experimental-induced asthma. As found for L-selectin, E-selectin/Ig chimeras can alter inflammatory cell recruitment (Mulligan et al., 1993). Though additional *in vivo* studies are needed, these preliminary studies show that E-selectin is important for the recruitment of myeloid cells to sites of inflammation and they correlate with the *in vitro* binding studies.

An important issue is whether inhibitors of E-selectin can alter lymphocyte migration. Analysis of memory lymphocytes will be difficult due to their small and variable numbers. Furthermore, the memory T cell/E-selectin interaction has only been shown in the human. The ruminant γ/δ T cells are the predominant lymphocyte in the blood of young ruminants, and they are found in very large numbers in tissues associated with increased E-selectin expression (see preceding). We have shown that the accumulation of γ/δ T cells in the inflamed dermis can be blocked with anti-E-selectin antibodies (Jutila, unpublished). This represents the first direct demonstration that E-selectin can serve as a vascular addressin for the localization of lymphocytes to the skin.

C. P-selectin

P-selectin was originally isolated from activated platelets. Antibodies that stained thrombin-activated but not unactivated platelets were used to immunoprecipitate a 140 kD surface glycoprotein. The molecule was later found in the Weibel-Palade bodies of endothelial cells and demonstrated to be quickly translocated (within minutes) to the cell surface following thombin or histamine stimulation (Berman et al., 1986; Stenberg et al., 1985; McEver et al., 1989; Bonfanti et al., 1989). The function of P-selectin was fully appreciated after cloning of its cDNA (see below) and determining its relationship to L- and E-selectin. Following the cloning studies, anti-P-selectin antibodies were shown to block activated platelet binding to myeloid cells, such as neutrophils and monocytes (Larsen et al., 1989, 1990). Human endothelial cells stimulated with histamine or thombin also support P-selectin-dependent adhesion of neutrophils (Geng et al., 1990; Lorant et al., 1993; Jones et al., 1993).

P-selectin is expressed by most endothelial cells *in vivo,* but, as indicated earlier, it is normally contained within storage granules of the cell. Conclusively demonstrating that surface expression of P-selectin correlates with influxes of leukocytes as shown for E-selectin is difficult to do because activated platelets, which also express P-selectin, line venules in sites of inflammation. Thus, potential surface expression of P-selectin could come from the endothelial cell or the platelet.

In vivo studies of inhibiting P-selectin have only recently been started. A number of abstracts at recent meetings have shown that inhibitors of P-selectin are effective in blocking acute inflammatory lung injury following systemic activation of complement (Mulligan and Ward, 1993). Seekamp et al. (1993) have shown that P-selectin is important in the early phases of reperfusion injury in the lung and skin. Soluble P-selectin/Ig chimeras also block inflammatory reactions in vivo (Mulligan et al., 1993). P-selectin knock-out mice have been generated in which leukocyte recruitment is impaired (Mayadas et al., 1993). Obviously additional analyses are needed, but these preliminary studies suggest that inhibitors of P-selectin block certain types of inflammatory events. This is not unexpected in light of the in vitro data which show that P-selectin mediates both leukocyte/endothelial-cell and leukocyte/platelet interactions. Indeed, it is likely that in certain inflammatory processes, inhibitors of P-selectin will be far more effective than inhibitors of E- and L-selectin.

III. SELECTINS MEDIATE LEUKOCYTE ADHESION UNDER FLOW

The initial interaction of leukocytes with the vascular endothelium (step 1 from the model presented in the introduction) occurs under considerable shear forces associated with the flow of blood. Inhibitors of selectins specifically block these events in vivo (Ley et al., 1991; von Andrian et al., 1991, 1992). A number of *in vitro* systems have been developed which reproduce conditions in the blood. Using these systems, it has been confirmed that the interactions that occur under flow are preferentially mediated by selectins, whereas static adhesion is mediated by leukocyte integrins and their counter-receptors (Butcher, 1991; Springer, 1990; Lawrence and Springer, 1991, 1993; Abbassi et al., 1993; Bargatze et al., 1994). Indeed, the original assay which was used in the characterization of L-selectin measured leukocyte/endothelial-cell interactions under constant rotation (Stamper and Woodruff, 1976). Recently, L-selectin was shown to require shear to mediate cell achesion (Finger et al., 1996). These findings provide direct functional support for the role of selectins in the first step of the leukocyte extravasation process.

IV. REGULATION OF SELECTIN EXPRESSION

The expression of all three selectins is uniquely regulated. E- and P-selectin are normally not expressed on unactivated, noninflamed endothelial cells or platelets. Expression of these protein occurs in response to inflammatory signals. In contrast, L-selectin is constitutively expressed by most leukocytes and does not require external, activating signals for function. Interestingly, L-selectin is rapidly down regulated from the cell surface following activation of the leukocyte, which is unlike any other leukocyte adhesion protein. Neutrophils isolated from an inflammatory site express little L-selectin, whereas cells in circulation express high levels of the antigen (Jutila et al., 1989). Treating neutrophils with chemotactic factors *in vitro* causes a rapid (within minutes) loss of L-selectin from the cell surface (Jutila et al., 1989; Kishimoto et al., 1989; Griffin et al., 1990), which is due to shedding of the surface molecule (Kishimoto et al., 1989). L-selectin shedding also occurs *in vivo*—high levels of shed L-selectin can be detected in blood (Palecanda et al., 1992; Schleiffenbaum et al., 1992). In contrast to the loss of L-selectin following activation, the expression and functional activity of beta-2 integrins, Mac-1 (CD11b/CD18), for example, increases (Kishimoto et al., 1989).

Since a large fragment of L-selectin, which is slightly smaller than the native molecule, can be isolated from the supernatant of activated cells, it is assumed that the release of L-selectin is due to proteolysis (Kishimoto et al., 1989; Jutila et al., 1991; Jung and Daily, 1990). Recently, insight has been gained into the region that is clipped, based on sequence analysis of the "stump" left after shedding and site-directed mutagensis analysis (Kahn et al., 1994; Migaki et al., 1995; Chen et al., 1995). From these studies, it has been concluded that L-selectin is clipped just outside of the transmembrane domain and that the distance between the transmembrane domain and the first SCR is more critical in regulating the proteolytic event than the amino acid sequences themselves.

The potential functional implications of the shedding of L-selectin during leukocyte extravasation have not been conclusively shown. Kishimoto (1991) and I (Jutila, 1992) have suggested that the shedding of L-selectin following receptor engagement may allow the leukocyte to break its tight bonds with the vascular endothelium and proceed with emigration into the underlying tissue. In support of this hypothesis, we have found that bovine γ/δ T cells do not efficiently shed L-selectin, which correlates with an inability of these cells to migrate into peripheral lymph nodes following their L-selectin-mediated binding to HEV (Walcheck and Jutila, 1994). Another possibility is that shedding may also contribute to the phenomenon of leukocyte rolling along the vasculature prior to permanent adhesion, which others have shown to be an L-selectin-dependent event (Ley et al., 1991; von Andrian et al., 1991, 1992). Finally, having the capacity to shed L-selectin allows the leukocyte, which may initially arrest on certain venules via L-selectin but does not proceed with extravasation (i.e., eosinophils in most sites of acute inflammation or lymphoid tissue), to re-enter the circulation. Developing means of inhibiting L-selectin shedding may allow testing of these hypotheses.

In addition to the short-term regulation of L-selectin involving proteolysis, the surface expressed of L-selectin is also regulated by changes in the expression of the L-selectin gene. Virgin lymphocytes are all thought to be L-selectin positive. Mitogen and/or antigen stimulation of these cells can lead to alterations in L-selectin gene transcription, including downregulation (Kaldijan et al., 1992; Kaldijan and Stoolman, 1993; Dailey, 1993). Since L-selectin-negative lymphocytes are readily found in circulation, the downregulation of L-selectin expression can be long term. Picker and colleagues (1993a) have found that most of the L-selectin-negative human lymphocytes in circulation are usually memory cells or recently activated lymphocytes.

Selective downregulation of L-selectin gene expression may also occur preferentially in mucosal-associated versus peripheral-lymphoid tissues *in vivo* (Picker et al., 1993a). The selective downregulation of L-selectin on cells that respond to antigen in mucosal tissue may help ensure that these cells preferentially migrate back to this tissue. (Tissue-specific homing and tissue-specific regulation of adhesion molecule expression are fascinating areas of study, but their review is beyond the intent of this chapter. The reader is directed to many excellent reviews of the subject, (e.g., Berg et al., 1989; Butcher, 1986; Picker and Butcher, 1992; Rosen, 1989.)

The expression of E-selectin on cytokine-stimulated endothelial cells requires *de novo* mRNA and protein synthesis. Kinetic analysis has shown that 2-4 h are needed for optimal E-selectin expression on the surface of endothelial cells following cytokine stimulation. E-selectin mRNA decline to basal levels by 24-48 h, which correlates with a decrease in surface protein expression (Bevilacqua et al., 1987, 1989). The NK-kB and AP-1 transcription factors appear to be involved in the inducible expression of E-selectin (Montgomery et al., 1991; Whelen et al., 1991). An interesting feature of E-selectin is that it is only expressed by endothelial cells. An important area of future research is identification of the regulatory sequences and mechanisms that control this cell-type-specific expression.

The down-regulation of E-selectin from the surface of activated endothelial cells likely involves internalization of the antigen, however, additional mechanisms may also be involved. Recent reports suggest that proteolytic cleavage and shedding of E-selectin takes place (Newman et al., 1993). This latter observation is similar to the regulation of L-selectin expression on leukocytes (see preceding). However, using E- and L-selectin cDNA-transfected mouse lymphoma cells, we have clearly shown that the same activation/proteolytic events that lead to L-selectin shedding have no effect on the expression of E-selectin (Jutila, unpublished). E-selectin expressed in mouse L-cells is also resistant to activation-induced shedding. Thus, if proteolytic cleavage and shedding of E-selectin takes place it is unlikely to be via the same mechanism involved in the regulation of L-selectin.

As mentioned above, P-selectin is contained within endothelial-cell and platelet storage granules. Kinetic studies *in vitro* show that P-selectin is rapidly (within minutes) expressed on the surface of platelets and endothelial cells following stimulation with acute inflammatory mediators. The surface expression of P-selectin appears to rapidly decay within 30 minutes after stimulation of endothelial cells *in vitro.* If the same kinetics of expression and downregulation occur *in vivo,* this would suggest that P-selectin's primary role is during the earliest phases of the acute inflammatory process. However, this hypothesis is complicated by recent observations that P-selectin can be found expressed at high levels on venules in sites of chronic inflammation. For example, Stoolman and colleagues (1992) have shown that P-selectin is expressed on venules in chronically inflamed synovium. Expression appears to be on the surface of the endothelial cells, but, as indicated earlier, this is difficult to conclusively show. If P-selectin can be expressed for long periods of time *in vivo,* elucidation of the regulatory events leading to this is an important goal.

V. STRUCTURE/FUNCTION ANALYSIS OF SELECTINS

As discussed in the introduction, cloning of selectin cDNAs revealed much of their potential structure and function. Based on sequence homology, selectins were found to have an N-terminal domain similar to mammalian C-type lectins (Siegelman et al., 1989; Lasky et al., 1989; Tedder et al., 1989; Camerini et al., 1989; Bevilacqua et al., 1989; Larsen et al., 1989; Johnston et al., 1989). Previous studies showed that L-selectin has lectin activity (reviewed in Rosen, 1990), but the finding of a lectin domain in E- and P-selectin was the first indication that carbohydrate recognition was involved in the function of these two molecules. In addition to having similar lectin domains, selectins also have an EGF domain, multiple SCRs, a transmembrane region, and a short cytoplasmic tail. The three selectins differ in the number of SCRs, and these regions exhibit the lowest levels of identity.

Using L-selectin/Ig chimeras, Rosen, Watson, and Lasky's group (reviewed in Lasky, 1992) have demonstrated that the lectin domain of L-selectin is needed for binding HEV and specific carbohydrates (see following). Most functional epitopes of L-selectin have been mapped to the lectin domain as well, but exceptions do exist (Kansas et al., 1991; Jutila et al., 1992; Siegelman et al., 1990; and see following).

Analysis of E- and P-selectin has also demonstrated the importance of the lectin domain (Berg et al., 1992; Foxall et al., 1992; Larsen et al., 1992; Moore et al., 1992). Recently, common sites in the lectin domains of both proteins have been identified that are involved in carbohydrate recognition and cell adhesion. Using both site-directed mutagenesis and antibody-mapping techniques, a small portion of the E- and P-selectin lectin domain was actually shown to be involved in the recognition event. Two loops adjacent to the antiparallel beta sheet appear to account for virtually all of the carbohydrate-binding capacity of both selectins which are important in cell/cell adhesion (Erbe et al., 1992, 1993). Evidence that E- and P-selectin have the same carbohydrate-binding region is interesting in light of previous reports showing that P-selectin binds a wider array of carbohydrates than E-selectin (Larsen et al., 1992). The nature of these carbohydrates is discussed ahead. From their work, Erbe and colleagues (1993) suggest that the different carbohydrate binding activities that others have measured are not important in cell/cell adhesion and may be in vitro artifacts. Those carbohydrate interactions that directly support cell/cell binding appear the same for both vascular selectins.

The SCRs may also contribute to the function of selectins, though their precise role is not well defined. Watson and colleagues (1991b) showed that the presence of the SCRs increases the binding activity of L-selectin. Results of studies using a new mAb (EL-246), which recognizes both E- and L-selectin, further support a potential role for the SCRs. EL-246 effectively blocks the function of L- and E-selectin in many different cell/cell adhesion assays (Jutila et al., 1992; Bargatze et al., 1994). Interestingly, EL-246 does not block the ability of L-selectin to bind soluble carbohydrate (Jutila et al., 1992). Preliminary *in vivo* analysis suggests that EL-246 is uniquely effective at blocking inflammation (Jutila, unpublished).

Though EL-246 blocks function, initial mapping studies have localized its epitope to the SCRs (Jutila et al., 1992). This was based on the selective staining of L- and P-selectin chimeras and crossblocking experiments with other selectin mAbs. EL-246 may block by disrupting an important conformation conferred by the SCRs that is required for selectin-mediated cell/cell interactions. Others have shown that oligomerization of L-selectin into dimers, trimers, and tetramers, which require the SCRs, may be important for increasing its affinity for ligand (Crommie and Rosen, 1992). EL-246 could prevent this from occurring. Alternatively, EL-246 may block a previously undescribed ligand interaction mediated by the SCRs. Clearly, more analysis must be done to define the precise location and nature of the EL-246 epitope and the role of the SCRs in facilitating ligand recognition.

The cytoplasmic tail of L-selectin also appears to be required for function. Deletion of a region conserved in all three selectins inhibited the ability of L-selectin to mediate lymphocyte binding to HEV and leukocyte rolling on mesenteric venules in vivo. Surprisingly, the lectin activity of L-selectin (binding of soluble carbohydrates) was unaffected by the deletion of the cytoplasmic tail region (Kansas et al., 1993). Kansas suggests that an interaction of the tail region with the cytoskeleton is important in maintaining an important functional conformation of L-selectin. This hypothesis is supported by the observation that cytochalasin B treatment, which disrupts the cytoskeleton, blocks the function of L-selectin on lymphocytes (Kansas et al., 1993).

In summary, the lectin domains of all three selectins are required for function via recognition of specific carbohydrate structures on target cells. Other regions of the molecules are also important in regulating function potentially through control of interactions with the cytoskeleton, maintaining appropriate conformation and orientation of the lectin domain, and, perhaps, via novel ligand interactions.

VI. CARBOHYDRATE LIGANDS FOR SELECTINS

As discussed earlier, structural analysis of selectins has demonstrated that their carbohydrate-binding lectin domains are required for function. Earlier studies showed that mannose-6-PO4 or mannose-6-PO4-rich polysaccharides, such as the phosphomannan PPME, bound L-selectin (reviewed in Rosen, 1990). These carbohydrates also block the ability of lymphocytes to bind peripheral-lymph-node HEV. The nature of the native carbohydrates expressed by endothelial cells that serve as ligands for L-selectin still has not been defined.

In contrast to L-selectin, much is known about the carbohydrate structures on myeloid cells bound by the vascular selectins. The first insight into the nature of these sugars came from expression cloning experiments in which transfection of a cDNA encoding a fucosyl transferase, necessary to produce the sialyl Lewis x (SLe^x; sialic acid alpha 2-3 galactose beta 1-4 {fucose alpha 1-3} N-acetyl glucosamine) structure, conferred P- and E-selectin binding in cells that normally did not bind these selectins (Lowe et al., 1990; Goelz et al., 1990). Additional analyses eventually showed that SLe^x on myeloid cells and the related SLe^a predominantly found on tumor cells, are directly bound by E-selectin (Phillips et al., 1990; Walz et al., 1990; Berg et al., 1992; Foxall et al., 1992; Larsen et al., 1992; Larkin et al., 1992). Fucose linked to the subterminal rather than to an internal N-acetylglucosamine is a requirement for binding, and the presence of sialic acid-3 linked to the terminal galactose of these carbohydrate structures substantially enhances the binding affinity (Larkin et al., 1992). 2,6 linked SLe^x has also been suggested to be an important ligand for P- but not E-selectin (Larsen et al., 1992), however, the mapping studies by Erbe, just outlined, showed that this interaction was not important in cell/cell adhesion. Thus, the native carbohydrate ligand on myeloid cells of importance in vivo is the 2,3 linked SLe^x structure.

Since SLe^x is normally not found on lymphoid cells, it is of considerable interest to define the carbohydrate structures expressed by memory lymphocytes that serve as ligands for E-selectin. The HECA 452 mAb may recognize these structures (Berg et al., 1991b; Picker et al., 1990, 1991b, 1993b). As discussed previously, expression of the HECA 452 epitope correlates with the ability to bind E-selectin. In some assays, HECA 452 blocks adhesion to E-selectin. Interestingly, HECA 452 recognizes SLe^x on neutrophils and sialylated carbohydrates on lymphocytes (Berg et al., 1991a, 1992). From these results, Berg suggests that HECA 452 recognizes a common structure and/or conformation of certain carbohydrate ligands for E-selectin. Furthermore, the lymphocyte carbohydrates are related to SLe^x.

Additional antigenically distinct carbohydrate ligands for E-selectin also exist. Antibodies that recognize SLe^x and the HECA 452 mAb do not stain bovine γ/δ T cells, which avidly bind E-selectin. However, sialylated carbohydrates are important in this interaction because neuraminidase treatment of the γ/δ T cell completely blocks its ability to bind E-selectin (Walcheck et al., 1993). These results show that multiple, antigenically distinct carbohydrate structures can serve as E-selectin ligands.

In recent studies comparing the carbohydrate binding specificities of recombinant L-, E-, and P-selectin, Berg and colleagues (1992) and Foxall and colleagues (1992) have shown that all three selectins can bind soluble SLe^x, though differences in affinity were noted. This result is not surprising considering the level of homology between the lectin domains of all three selectins, but their *in vivo* relevance is unknown.

VII. HIGH-AFFINITY PROTEIN LIGANDS FOR SELECTINS

The carbohydrates discussed above represent important but incomplete representations of the entire ligands for the selectins. Structures on the surface of target cells must present the sugars in appropriate conformation and concentration to get high-affinity cell/cell interactions. It is conceivable that lipid or protein could serve as the scaffolding component of the ligands. Indeed, indirect evidence

Table 1. Selectin Glycoprotein Ligands

Glycoprotein	*Ligand for*	*Mucin-like*	*Reference*
PNAd	L-selectin	?	Berg et al. (1991a)
GlyCAM-1	L-selectin	Yes	Lasky et al. (1992a)
CD34	L-selectin	Yes	Baumheuter et al. (1993)
MAdCAM-1[a]	L-selectin	Yes	Briskin et al. (1993)
E-selectin	L-selectin	No	Picker et al. (1991a)
P-selectin	L-selectin	No	Picker et al. (1991a)
150 kD gp	E-selectin	?	Levinovitz et al. (1993)
250 kD gp[b]	E-selectin	?	Walcheck et al. (1993)
120 kD gp	E-selectin	Yes	Sako et al. (1993)
L-selectin	E-selectin	No	Picker et al. (1991a)
120 kD gp	P-selectin	Yes	Sako et al. (1993)
L-selectin	P-selectin	No	Picker et al. (1991a)

Notes: [a] MAdCAM-1 was originally defined as the mucosal HEV vascular addressin, which is bound by $\alpha4/\beta7$ on lymphocytes. Some forms of MAdCAM-1 express the PNAd epitope and bind L-selectin.
[b] All of the E- and P-selectin ligands are expressed by neutrophils except the 250 kD molecule, which is expressed by γ/δ T cells.

for a role of lipids on the surface of neutrophils in binding soluble E-selectin has been reported (Larsen et al., 1992). Though glycolipids may be important for the interaction of some cells with selectins in certain instances, affinity isolation techniques, protease studies, antibody blocking experiments, and cDNA transfection experiments suggest that a small number of cell-surface glycoproteins may represent the key ligands for the selectins. These ligands are all large molecular weight surface glycoproteins, some of which have domains rich in carbohydrate, similar to mucin-like molecules (Shimizu and Shaw, 1993). Table 1 lists putative glycoprotein receptors for selectins that have been identified. In the section that follows, the nature of the defined mucin-like selectin ligands will be described.

A. Glycoprotein Ligands for L-selectin

As outlined, HEV in peripheral lymph nodes constitutively express ligands for L-selectin. MECA 79 is a rat monoclonal antibody that reacts with peripheral lymph node HEV and blocks lymphocyte adhesion to these venules (Streeter et al., 1988). Lower reactivity of MECA 79 is seen with HEV in mucosal lymphoid tissues. Injection of MECA 79 into mice blocks lymphocyte homing to peripheral lymph nodes, but has minimal effects on the accumulation of the cells into the gut. Lymphocyte adhesion to affinity-purified MECA 79 antigen is blocked by treatment of the lymphocyte with anti-L-selectin antibodies, suggesting that L-selectin and the MECA 79 antigen interact (Berg et al., 1991a). Western blot analysis of the MECA 79 antigen reveals multiple molecular-weight glycoproteins, with the predominant species having a molecular weight of 90 kD (Berg et al., 1991a). Analyses have not been done to determine if the 90 kD MECA 79 reactivity antigen represents the predominant L-selectin ligand on peripheral lymph node HEV. The MECA 79 antigen has been termed the peripheral lymph node vascular addressin or PNAd.

Lasky and colleagues (1992) used a recombinant immunoglobulin/L-selectin chimera to purify a 50 kD glycoprotein from peripheral-lymph-node HEV, sequenced the isolated molecule, and identified its cDNA. They termed the 50 kD molecule GlyCAM-1. GlyCAM-1 binds L-selectin, which can be inhibited by EDTA, or neuraminidase. The distribution of GlyCAM-1 correlates with the HEV binding mediated by L-selectin (predominantly in lymph nodes) and MECA 79 reacts with the native molecule (Imai et al., 1991; Lasky et al., 1992). Sequence analysis of the GlyCAM-1 cDNA predicts two regions that are heavily glycosylated with O-linked sugars, characteristic of mucin-like molecules (Lasky et al., 1992; Lasky, 1992). It has been proposed that the high concentration of carbohydrate in GlyCAM-1 may facilitate its recognition by L-selectin. Imai and colleagues (1993) have shown that sulphation of GlyCAM-1 is also important in its interaction with L-selectin. Though soluble GlyCAM-1 can interact with L-selectin, it is not clear if it represents the predominant molecule that regulates the binding of L-selectin-positive lymphocytes to HEV. To date, no function-blocking mAb has been directly raised against GlyCAM-1 (reviewed in Lasky, 1992).

The L-selectin/Ig chimera also immunoprecipitates a 90 kD molecule from mouse HEV. Recently, Baumhueter and colleagues (1993) found that a glycoform of CD34 that is expressed by endothelial cells represents this 90 kD ligand. As found for GlyCAM-1, CD34 exhibits similarities to mucins. MECA 79 also reacts with CD34.

Lymphocyte traffic through mucosal lymphoid tissue is regulated by an endothelial-cell adhesion molecule distinct from PNAd. MECA 367 recognizes the mucosal adhesion protein (now termed

MAdCAM-1) and its cDNA has recently been cloned (Briskin et al., 1993). Sequence analysis shows that MAdCAM-1 contains a mucin domain similar to that found in GlyCAM-1 (Briskin et al., 1993). Berg and colleagues (1993) have recently shown that MAdCAM-1 isolated from mesentery lymph nodes is modified by the MECA 79 eptitope and can support L-selectin-dependent adhesion and rolling of lymphocytes. Thus, MADCam-1 represents another identified mucin which is expressed by lymphoid tissue HEV that serves as a ligand for L-selectin.

The nature of the L-selectin ligand on cytokine-activated endothelial cells is still not completely defined. CD34 is expressed by endothelium *in vivo* and *in vitro* (Baumhueter et al., 1993), but it has not been shown to be involved in leukocyte adhesion to inflamed endothelium. We showed that E-selectin on activated endothelial cells may be a counter-receptor for L-selectin (Kishimoto et al., 1990b). Neutrophils adhere to E-selectin cDNA-transfected L-cells, which can be blocked by treatment of the leukocyte with anti-L-selectin antibodies (Kishimoto et al., 1990b). L-selectin isolated from neutrophils but not lymphocytes supports the adhesion of E-selectin transfectants (Picker et al., 1991a). We (Bargatze et al., 1993) and others (Abbassi et al., 1993) have shown that neutrophil rolling on cytokine-activated endothelial cells expressing E-selectin or on the E-selectin transfectants can be blocked with anti-L-selectin antibodies. Therefore, under certain conditions, E-selectin may support L-selectin-dependent adhesion. Spertini and colleagues (1991a) suggest that other undefined L-selectin ligands exist on cytokine-activated endothelial cells. It is possible that these other ligands are related to CD34, but this remains to be determined (Baumhueter et al., 1993).

Recently, we showed that once neutrophils actively adhere to the vascular endothelium during the extravazation process they can support continued rolling of newly arriving neutrophils. Anti-L-selectin antibodies completelyl block this leukocyte on leukocyte rolling (Bargatze et al., 1994). Immobilized gamma/delta T cells also support leukocyte-on-leukocyte rolling, which is completely dependent on L-selectin (Jutila and Kurk, 1996). The L-selectin ligand on leukocytes is not fully characterized, but is a mucin which expresses sialylated carbohydrates (Jutila and Kurk, 1996). PSGL-1 (see following) could serve as the leukocyte L-selectin ligand.

B. Glycoprotein Ligands for E-selectin

In the past few years, conflicting reports have been published on the importance of protein in the ligand for E-selectin on leukocytes. In some settings, proteases have no effect on neutrophil binding of E-selectin (Larsen et al., 1992), whereas in others they effectively block the interaction (Picker et al., 1991a). Variations in the conditions of the assays and the cell types which are examined (normal leukocytes versus tumor cell lines) may account for some of these differences. Incomplete proteolysis may also have lead to a lack of a blocking effect in some of the analyses. Picker and colleagues (1991a) showed the chymotrypsin treatment effectively blocks neutrophil binding to E-selectin. In our studies of γ/δ T cells, we have found that chymotrypsin or trypsin treatment of the lymphocyte effectively blocks their ability to bind E-selectin (Walcheck et al. 1993).

In addition to the blocking effects of proteases, direct evidence for glycoprotein ligands for E-selectin has also been obtained. As discussed, L-selectin on neutrophils can, under certain conditions, serve as a receptor for E-selectin (Picker et al., 1991a). However, L-selectin is not the only receptor for E-selectin nor has it been conclusively shown to be the predominant receptor *in vivo*. Variants of the HL-60 neutrophilic cell line exist that avidly bind E-selectin, but don't express L-selectin (Larsen et al., 1992). Further, using an immunoglobulin/E-selectin chimera, Vestweber's group (Levinovitz et al., 1993) identified a predominant 150 kD glycoprotein mouse neutrophil detergent lysate that is bound by E-selectin. Interestingly, L-selectin was not isolated from the lysates by the chimera, which may suggest that the L-selectin/E-selectin interaction does not readily occur in solution. Preliminary biochemical analyses indicate that the 150 kD molecule is heavily glycosylated. Recently, a cDNA was isolated fo the 150 kD ligand and the gene product called E-selectin ligand-1 (ESL-1; Steegmaier et al., 1995).

As just discussed, L-selectin on neutrophils can serve as a receptor for E-selectin, but a similar role for lymphocyte L-selectin does not occur (Picker et al., 1991a). γ/δ T cells, which avidly bind E-selectin, express very high levels of L-selectin (Walcheck and Jutila, 1994), but anti-L-selectin antibodies or treatments that specifically remove L-selectin from the lymphocyte surface have no effect on binding (Walcheck et al., 1993). The difference in the binding of L-selectin from neutrophils and lymphocytes to E-selectin may be related to different post-translational modifications of the molecules. Picker and colleagues (1991a) showed that neutrophil L-selectin is decorated with sLeX (a carbohydrate bound by E-selectin), whereas the lymphocyte molecule is not.

As indicated, protease treatment of the γ/δ T cell blocks its ability to bind E-selectin. Since L-selectin clearly is not the ligand controlling this interaction, we have attempted to identify a glycoprotein receptor on γ/δ T cells using native E-selectin as an affinity reagent. E-selectin was purified from detergent lysates of E-selectin cDNA-transfected mouse L-cells and bound to a sepharose column using a nonblocking anti-E-selectin mAb. γ/δ T cell detergent lysates were sequentially passed over control preclearing columns and the E-selectin column. Material bound to the E-selectin column was eluted by addition of EDTA, which reverses selectin-dependent interactions. Fractions were collected from the column, run on an 8% SDS-PAGE gel, and protein was revealed by silver staining. Using these procedures, we have specifically isolated a 250 kD (under nonreducing conditions) glycoprotein that exhibits all of the characteristics of a ligand for E-selectin: it is only expressed by cells that bind E-selectin; EDTA reverses its interaction with E-selectin; and it is modified by sialic acid. Furthermore, removal of sialic acid inhibits the capacity of the 250 kD glycoprotein to bind E-selectin (Walcheck et al., 1993). Current analyses suggest that the 250 kD molecule is heavily glycosylated, which is a recurring feature of defined selectin ligands.

CD45 has recently been proposed as a potential selectin ligand because it is a cell surface glycoprotein which has a mucin-like domain (Shimizu and Shaw, 1993). CD45 exist in many different isoforms on leukocytes. The molecular weights of some of these isoforms are over 200 kD. Thus, it is possible that the 250 kD glycoprotein we have isolated from γ/δ T cells is a CD45 species. We are currently attempting to determine this.

C. Glycoprotein Ligands for P-selectin

As seen in some studies of E-selectin, protease treatment of neutrophils effectively blocks their ability to bind P-selectin (Larsen et al., 1992; Moore et al., 1992). Picker and colleagues (1991a) provide preliminary evidence that L-selectin on neutrophils may participate in binding of P-selectin, as described above for E-selectin. However, these data are less convincing, since anti-L-selectin mAbs have little effect on the ability of neutrophils to bind soluble P-selectin (Moore et al., 1992). Moore (1992) used isolated P-selectin bound to sepharose as an affinity reagent to identify putative glycoprotein ligands on neutrophils. Using a purification procedure we based ours on, these authors identified a 250 kD (under nonreducing conditions) glycoprotein on neutrophils that is specifically bound by P-selectin. Under reducing conditions, the molecule runs around 120 kD, suggesting that the nonreduced form represents a dimer of the 120 kD molecule. O-linked sugars, as predicted for the L-selectin ligand GlyCAM-1, appear to contribute significantly to the overall molecular weight of the ligand (Norgard et al., 1993). Recently, a cDNA for the P-selectin glycoprotein ligand-1 (PSGL-1) on neutrophils has been isolated. As found for GlyCAM-1, the sequence of the neutrophil ligand predicts a mucin region rich in O-linked carbohydrates (Sako et al., 1993). Antibodies generated against the recombinant molecule stain neutrophils and react with the 120 kD glycoprotein purified by the P-selectin affinity column. Recently, anti-PSGL-1 mAb were shown to block neutrophil rolling on P-selectin (Norman et al., 1995).

D. Speculation of the Existence of a Family of High Affinity Selectin Ligands

As briefly outlined, evidence has accumulated for the existence of a group of novel highly glycosylated glycoprotein ligands for all three selectins (Shimizu and Shaw, 1993). A common feature of the best defined examples of these molecules is their isolation from target-cell detergent lysates by ligand-affinity techniques using purified selectins or recombinant selectin/immunoglobulin chimeras. These large molecular weight glycoproteins serve to concentrate sugars in appropriate conformation and present them in a fashion that supports cell-cell interactions under shear forces associated with blood flow. Though the evidence supporting an important role for glycoproteins as selectin ligands is compelling, *in vivo* studies of their importance are lacking. Specific inhibitors as well as gene knock-out mice need to be developed to examine their function in the animal.

VIII. FUTURE RESEARCH DIRECTIONS

The study of selectins has contributed significantly to our understanding of the extravasation process. We know much about their expression and function in vivo, but there is still a lot that we don't know. As previously outlined, a critical area of future research on selectins is in molecular characterization of the high-affinity glycoprotein ligands; foremost in these experiments is the generation of function-blocking antibodies. Other important areas for future study include (1) elucidation of the regulation of selectin gene expression; (2) characterization of the signalling events required for L-selectin shedding; (3) determination of any role for protein/protein interactions between selectins and their ligands; and (4) additional in vivo analysis. This latter point is particularly relevant to the generation of new, anti-inflammatory therapeutics to treat diseases, such as arthritis, psoriasis, sepsis, and reperfusion injury.

ACKNOWLEDGMENT

The expert technical assistance of Sandy Kurk, Gayle Watts, and Kathryn Jutila was instrumental in the generation of much of the data reviewed here and is greatly appreciated, as was the contribution of former coworkers and collaborators: Rupert Hallmann, Frans Kroese, Ellen Berg, T. Kei Kishimoto, Louis Picker, and especially Eugene Butcher. The role of graduate students, Aiyappa Palecanda, Bruce Walcheck, and Rob Bargatze in generating much of the new data summarized here is also acknowledged. The efforts of Dana Hoover in the preparation of the manuscript are greatly appreciated. Parts of these studies were funded by grants from the USDA (CRSR-90-01666), Pardee Research Foundation, American Cancer Society (ACS CD476), and the Montana Agricultural Experiment Station.

REFERENCES

Abbassi, O., Kishimoto, T.K., Lane, C.L., McIntyre, L.V., & Smith, C.W. (1993). E-selectin supports neutrophil rolling under conditions of flow. J. Clin. Invest. 92, 2719-2730.

Abbassi, O., Lane, C.L., Krater, S., Kishimoto, T.K., Anderson, D.C., McIntire, L.V., & Smith, C.W. (1991). Canine neutrophil margination mediated by lectin adhesion molecule-1 *in vitro*. J. Immunol. 147, 2107.

Allison, J.P., & Havran, W.L. (1991). The immunobiology of T cells with invariant γ/δ antigen. Ann. Rev. Immunol. 9, 679-706.

Arbones, M.L., Ord, D.C., Ley, K., Ratech, H., Maynard-Curry, C., Otten, G., Capon, D.J., & Tedder, T.F. (1994). Lymphocyte homing and leukocyte rolling and migration are impaired in L-selectin (CD62L) deficient mice. Immunity 1, 247.

Bargatze, R.F., Kurk, S., Watts, G., Kishimoto, T.K., Speer, C.A., & Jutila, M.A. (1994). *In vivo* and *in vitro* functional examination of a conserved epitope of L- and E-selectin crucial for leukocyte-endothelial cell interactions. J. Immunol. 152, 5814-5825.

Baumhueter, S., Singer, M.S., Henzel, W., Hemmerich, S., Renz, M., Rosen, S.D., & Lasky, L.A. (1993). Binding of L-selectin to the vascular sialomucin CD34. Science 262, 436-438.

Berg, E.L., Goldstein, L.A., Jutila, M.A., Nakache, M., Picker, L.J., Streeter, P.R., Wu, N.W., Zhou, D., & Butcher, E.C. (1989). Lymphocyte homing receptors and vascular addressins: Cell adhesion molecules that direct lymphocyte traffic. Immunol. Rev. 108, 5-18.

Berg, E.L., McEvoy, L.M., Berlin, C., Bargatze, R.F., & Butcher, E.C. (1993). L-selectin-mediated lymphocyte rolling on MAdCAM-1. Nature 366, 695-698.

Berg, E.L., Robinson, M.K., Warnock, R.A., & Butcher, E.C. (1991a). The human peripheral lymph node vascular addressin is a ligand for LECAM-1, the peripheral lymph node homing receptor. J. Cell. Biol. 114, 343-349.

Berg, E.L., Yoshino, T., Rott, L.S., Robinson, M.K., Warnock, R.A., Kishimoto, T.K., Picker, L.J., & Butcher, E.C. (1991b). The cutaneous lymphocyte antigen is a skin lymphocyte homing receptor for the vascular lectin endothelial cell-leukocyte adhesion molecule 1. J. Exp. Med. 174, 1461-1466.

Berg, E.L., Magnani, J., Warnock, R.A., Robinson, M.K., & Butcher, E.C. (1992). Comparison of L-selectin and E-selectin ligand specifications: The L-selectin can bind the E-selectin ligands sialyl Lex and sialyl Lea. Biochem. and Biophys. Res. Comm. 184, 1048-1055.

Berman, C.L., Yeo, E.L., Wencel-Drake, J.D., Furie, B.C., Ginsberg, M.H., & Furie, B. (1986). A platelet alpha granule membrane protein that is associated with the plasma membrane after activation. J. Clin. Invest. 78, 130-137.

Bevilacqua, M., Butcher, E., Furie, B., Furie, B., Gallatin, M., Gimbrone, M., Harlan, J., Kishimoto, K., Lasky, L., McEver, R., Paulson, J., Rosen, S., Seed, B., Siegelman, M., Springer, T., Stoolman, K., Tedder, T., Varki, A., Wagner, D., Weissman, L., & Zimmerman, G. (1991). Selectins: A family of adhesion receptors. Cell 67, 233.

Bevilacqua, M.P., Pober, J.S., Mendrick, D.L., Cotran, R.S., & Gimbrone, M.A., Jr. (1987). Identification of an inducible endothelial-leukocyte adhesion molecule. Proc. Natl. Acad. Sci. USA 84, 9238-9242.

Bevilacqua, M.P., Stengelin, S., Gimbrone, M., & Seed, B. (1989). Endothelial leukocyte adhesion molecule 1: An inducible receptor for neutrophils related to complement regulatory proteins and lectins. Science (Wash. D.C.) 243, 1160-1172.

Bonfanti, R., Furie, B.C., Furie, B., & Wagner, D.D. (1989). PADGEM (GMP140) is a component of Weibel-Palade bodies of human endothelial cells. Blood 73, 1109-1112.

Briskin, M.J., McEvoy, L.M., & Butcher, E.C. (1993). MAdCAM-1 has homology to immunoglobulin and mucin-like adhesion receptors and to IgA1. Nature 363, 461-464.

Butcher, E.C. (1986). The regulation of lymphocyte traffic. Curr. Top. Micro. Immunol. 128, 85-122.

Butcher, E.C. (1991). Leukocyte-endothelial cell recognition: Three (or more) steps to specificity and diversity. Cell 67, 1033-1036.

Camerini, D., James, S.P., Stamenkovic, I., & Seed, B. (1989). Leu-8/TQ1 is the human equivalent of the MEL-14 lymph node homing receptor. Nature (London) 342, 78-80.

Chen, A., Engel, P., & Tedder, T.F. (1995). Structural requirements regulate endoproteolytic release of the L-selectin (CD62L) adhesion receptor from the cell surface of leukocytes. J. Exp. Med. 182, 519-530.

Cotran, R.S. (1987). New roles for the endothelium in inflammation and immunity. Am. J. Path. 129, 407-413.

Crommie, D., & Rosen, S.D. (1992). The lymph node homing receptor (L-selectin) forms oligomers. FASEB J. 6, A1191.

Dailey, M.O. (1993). The selectin family of cell adhesion molecules. In: *Lymphocyte Adhesion Molecules,* (Shimizu, Y., Ed.). R.G. Landes Company, p. 75-104.

Erbe, D.V., Wolitzky, B.A., Presta, L.G., Norton, C.R., Ramos, R.J., Burns, D.K., Rumberger, J.M., Narasinga Rao, B.N., Foxall, C., Brandley, B.K., & Lasky, L.A. (1992). Identification of an E-selectin region critical for carbohydrate recognition and cell adhesion. J. Cell. Biol. 119, 215-227.

Erbe, D.V., Watson, S.R., Presta, L.G., Wolitzky, B.A., Foxall, C., Brandley, B.K., & Lasky, L.A. (1993). P- and E-selectin use common sites for carbohydrate ligand recognition and cell adhesion. J. Cell. Biol. 120, 1227-1235.

Finger, E.B., Puri, K.D., Alon, R., Lawrence, M.B., von Andrian, U.H., & Springer, T.A. (1996). Adhesion through L-selectin requires a threshold hydrodynamic shear. Nature 379, 266-269.

Foxall, C., Watson, S.R., Dowbenko, D., Fennie, C., Lasky, L.A., Kiso, M., Hasegawa, A., Asa, D., & Brandley, B.K. (1992). The three members of the selectin receptor family recognize a common carbohydrate epitope, the sialyl Lewisx Oligosaccharide. J. Cell Biol. 117, 895-902.

Gallatin, W.M., Weisman, I.L., & Butcher, E.C. (1983). A cell surface molecule involved in organ-specific homing of lymphocytes. Nature 304, 30-34.

Geng, J.G., Bevilacqua, M.P., Moore, K.L., McIntyre, T.M., Prescott, S.M., Kim, J.M., Bliss, G.A., Zimmerman, G.A., & McEver, R.P. (1990). Rapid neutrophil adhesion to activated endothelium mediated by GMP-140. Nature (London) 343, 757-760.

Goelz, S.E., Hession, C., Goff, D., Griffiths, B., Tizard, R., Newman, B., Chi, R.G., & Lobb, R. (1990). ELFT: A gene that directs the expression of an ELAM-1 ligand. Cell 63, 1349-1356.

Graber, N., Gopal, T.V., Wilson, D., Beall, L.D., Polte, T., & Newman, W. (1990). T cells bind to cytokine-activated endothelial cells via a novel inducible sialoglycoprotein and endothelial leukocyte adhesion molecule-1. J. Immunol. 145, 819-830.

Griffin, J.D., Spertini, O., Ernst, T.J., Belvin, M.P., Levine, H.B., Kanakura, Y., & Tedder, T.F. (1990). Granulocyte-macrophage colony-stimulating factor and other cytokines regulate surface expression of the leukocyte adhesion molecule-1 on human neutrophils, monocytes, and their precursors. J. Immunol. 145, 576-584.

Gundel, R.H., Wegner, C.D., Torcellini, C.A., Clarke, C.C., Haynes, N., Rothlein, R., Smith, C.W., & Letts, L.G. (1991). Endothelial leukocyte adhesion molecule-1 mediates antigen-induced acute airway inflammation and late-phase airway obstruction in monkeys. J. Clin. Invest. 88, 1407-1411.

Hallmann, R., Jutila, M.A., Smith, C.E., Anderson, D.C., Kishimoto, T.K., & Butcher, E.C. (1991). The peripheral lymph node homing receptor, LECAM-1, is involved in CD18-independent adhesion of human neutrophils to endothelium. Biochem. Biophys. Res. Comm. 174, 236-243.

Imai, Y., Singer, M.S., Fennie, C., Lasky, L.A., & Rosen, S.D. (1991). Identification of a carbohydrate-based endothelial ligand for a lymphocyte homing receptor. J. Cell. Biol. 113, 1213-1221.

Imai, Y., Lasky, L.A., & Rosen, S.D. (1993). Sulphation requirement for GlyCAM-1, an endothelial ligand for L-selectin. Nature 361, 555-557.

Johnston, G.I., Cook, R.G., & McEver, R.P. (1989). Cloning of GMP-140, a granule membrane protein of platelets and the endothelium; sequence similarity to proteins involved in cell adhesion and inflammation. Cell 56, 1033-1044.

Jones, D.A., Abbassi, O., McIntire, L.V., McEver, R.P., & Smith, C.W. (1993). P-selectin mediates neutrophil rolling on histamine-stimulated endothelial cells. Biophysical. J. 65, 1560-1569.

Jung, T.M., & Dailey, M.O. (1990). Rapid modulation of homing receptors (gp90MEL-14) induced by activators of protein kinase C: Receptor shedding due to accelerated proteolytic cleavage at the cell surface. J. Immunol. 144, 3130-3136.

Jutila, M.A. (1992). Leukocyte traffic to sites of inflammation. Act. Path. Micro.Immun. Scan. 100, 191.

Jutila, M.A., & Kurk, S. (1996). Analysis of bovine g/d T cell interactions with E-, P-, and L-selectin: Characterization of lymphocyte on lymphocyte rolling and the effects of O-glycoprotease. J. Immunol. 156, 289-296.

Jutila, M.A., Rott, L., Berg, E.L., & Butcher, E.C. (1989). Function and regulation of the neutrophil MEL-14 antigen *in vivo:* Comparison with LFA-1 and Mac-1. J. Immunol. 143, 3318-3324.

Jutila, M.A., Lewinsohn, D.M., Lakey Berg, E., & Butcher, E.C. (1990). Homing receptors in lymphocyte, monocyte, and neutrophil interaction with endothelium. In: *Structure and Function of Molecules Involved in Leukocyte Adhesion* (Rosenthal, A.S., Springer, T.A., Anderson, D.C., & Rothlein, R., Ed.). Springer-Verlag, New York, p. 227.

Jutila, M.A., Kishimoto, T.K., & Finken, M. (1991). Low dose Chymotrypsin treatment inhibits neutrophil migration into sites of inflammation *in vivo:* Effects on Mac-1 and Mel-14 adhesion protein expression and function. Cell. Immunol. 132, 201-214.

Jutila, M.A., Watts, G., Walcheck, B., & Kansas, G. (1992). Characterization of a functionally important and evolutionarily well-conserved epitope encoded by the short concensus repeats of E-selectin and L-selectin. J. Exp. Med. 175, 1565-1573.

Kahn, J., Ingraham, R.H., Shirley, F., Migaki, G.I., & Kishimoto, T.K. (1994). Membrane proximal cleavage of L-selectin: Identification of the cleavage site and a 6-kD transmembrane peptide fragment of L-selectin. J. Cell Biol. 125, 461-470.

Kaldjian, E.P., & Stoolman, L.M. (1993). Lymphocyte recirculation and recruitment. In: *Lymphocyte Adhesion Molecules* (Shimizu, Y., Ed.). R.G. Landes Company, p. 151-172.

Kaldijan, E.P., Reddy, B.P., & Stoolman, L.M. (1992). Transcriptional regulation of L-selectin expression in normal and malignant T cells. FASEB J. 6, A5516.

Kansas, G.S., Ley, K., Munro, J.M., & Tedder, T.F. (1993). Regulation of leukocyte rolling and adhesion to high endothelial venules through the cytoplasmic domain of L-selectin. J. Exp. Med. 177, 833-838.

Kansas, G.S., Spertini, O., Stoolman, L.M., & Tedder, T.F. (1991). Molecular mapping of functional domains of the leukocyte receptor for endothelium, LAM-1. J. Cell. Biol. 114, 351-358.

Kishimoto, T.K. (1991). A dynamic model for neutrophil localization to inflammatory sites. J. NIH Res. 3, 75-77.

Kishimoto, T.K., Jutila, M.A., Berg, E.L., & Butcher, E.C. (1989). Neutrophil Mac-1 and Mel-14 adhesion proteins are inversely regulated by chemotactic factors. Science (Wash. D.C.). 245, 1238-1241.

Kishimoto, T.K., Jutila, M.A., & Butcher, E.C. (1990a). Identification of a human peripheral lymph node homing receptor: A rapidly down-regulated adhesion molecule. Proc. Natl. Acad. Sci. USA 87, 2244-2248.

Kishimoto, T.K., Warnock, R.A., Jutila, M.A., Butcher, E.C., Anderson, D.C., & Smith, C.W. (1990b). Antibodies against human neutrophil LECAM-1 (LAM-1/Leu-8 antigen/DREG56) and endothelial cell ELAM-1 inhibit a common CD18-independent adhesion pathway in vitro. Blood 78, 805-811.

Larkin, M., Ahern, T.J., Stoll, M.S., Shaffer, M., Sako, D., O'Brien, J., Yuen, C.-T., Lawson, A.M., Childs, R.A., Barone, K.M., Langer-Safer, P.R., Hasegawa, A., Kiso, M., Larsen, G.R., & Feizi, T. (1992). Spectrum of sialylated and nonsialylated Fuco-oligosaccharides bound by the endothelial-leukocyte adhesion molecule E-selectin. J. Biol. Chem. 267, 13661-13668.

Larsen, E., Celi, A., Gilbert, G.E., Furie, B.C., Erban, J.K., Bonfanti, R., Wagner, D.D., & Furie, B. (1989). PADGEM protein: A receptor that mediates the interaction of activated platelets with neutrophils and monocytes. Cell 59, 305-312.

Larsen, E., Palabrica, T., Sajer, S., Gilbert, G.E., Wagner, D.D., Furie, B.C., & Furie, B. (1990). PADGEM-dependent adhesion of platelets to monocytes and neutrophils mediated by a lineage-specific carbohydrate, LNF III (CD15). Cell 63, 467-474.

Larsen, G.R., Sako, D., Ahern, T.J., Shaffer, M., Erban, J., Sajer, S.A., Gibson, R.M., Wagner, D.D., Furie, B.C., & Furie, B. (1992). P-selectin and E-selecin: Distinct but overlapping leukocyte ligand specificities. J. Biol. Chem. 267, 11104-11110.

Lasky, L.A., Singer, M.S., Dowbenko, D., Imai, Y., Henzel, W., Grimley, C., Fennie, C., Gillet, N., Watson, S., & Rosen, S.D. (1992). An endothelial ligand for L-selectin is a novel mucin-like molecule. Cell 69, 927-938.

Lasky, L.A. (1992). Selectins: Interpreters of cell-specific carbohydrate information during inflammation. Science (Wash D.C.) 258, 964-969.

Lasky, L.A., Singer, M.S., Yednock, T.A., Dowbenko, D., Fennie, C., Rodriguez, H., Nguyen, T., Stachel, S., & Rosen, S.D. (1989). Cloning of a lymphocyte homing receptor reveals a lectin domain. Cell 56, 1045-1055.

Lawrence, M.A., & Springer, T.A. (1991). Leukocytes roll on a selectin at physiological flow rates: Distinct from and prerequisite for adhesion through integrins. Cell 65, 854-873.

Lawrence, M.A., & Springer, T.A. (1993). Neutrophils roll on E-selecin. J. Immunol. 151, 6338-6346.

Leung, D.Y.M., Pober, J.S., & Cotran, R.S. (1991). Expression of endothelial-leukocyte adhesion molecule-1 in elicited late phase allergic reactions. J. Clin. Invest. 87, 1805-1809.

Levinovitz, A., Muhlhoff, J., Isenmann, S., & Vestweber, D. (1993). Identification of a glycoprotein ligand for E-selectin on mouse myeloid cells. J. Cell. Biol. 121, 449-459.

Lewinsohn, D.M., Bargatze, R.F., & Butcher, E.C. (1987). Leukocyte-Endothelial cell recognition: Evidence of a common molecular mechanism shared by neutrophils, lymphocytes and other leukocytes. J. Immunol. 138, 4313-4321.

Ley, K., Gaehtgens, P., Fennie, C., Singer, M.S., Lasky, L.A., & Rosen, S.D. (1991). Lecin-like adhesion molecule-1 mediates leukocyte rolling in mesenteric venules in vivo. Blood 77, 2553-2555.

Lorant, D.E., Topham, M.K., Whatley, R.E., McEver, R.P., McIntyre, T.M., Prescott, S.M., & Zimmerman, G.A. (1993). Inflammatory roles of P-selectin. J. Clin. Invest. 92, 559-570.

Lowe, J.B., Stoolman, L.M., Nair, R.P., Larsen, R.D., Berhend, T.L., & Marks, R.M. (1990). ELAM-1 dependent cell adhesion to vascular endothelium determined by a transfected human fucosyltransferase cDNA. Cell 63, 475-484.

Mackay, C.R. (1991). Skin-seeking memory T cells. Nature (London) 349, 737-738.

Mayadas, T.N., Johnson, R.C., Rayburn, H., Hynes, R.O., & Wagner, D.D. (1993). Leukocyte rolling and extravasation are severely compromised in P selectin-deficient mice. Cell 74, 541-554.

McEver, R.P., Beckstead, J.H., Moore, K.L., Marshall-Carlson, L., & Bainton, D.F. (1989). GMP-140, a platelet alpha-granule membrane protein, is also synthesized by vascular endothelial cells and is localized in Weibel-Palade bodies. J. Clin. Invest. 84, 92-99.

Migaki, G.I., Kahn, J., & Kishimoto, T.K. (1995). Mutational analysis of the membrane-proximal cleavage site of L-selectin: Relaxed sequence specificity surrounding the cleavage site. J. Exp. Med. 182, 549-557.

Montgomery, K.F., Osborn, L., Hession, C., Tizard, R., Goff, D., Vassallo, C., Tarr, P.I., Bomsztyk, K., Lobb, R., & Harlan, J.M. (1991). Activation of endothelial-leukocyte adhesion molecule 1 (ELAM-1) gene transcription. Proc. Natl. Acad. Sci. USA 88, 6523-6527.

Moore, K.L., Stults, N.L., Diaz, S., Smith, D.F., Cummings, R.D., & Varki, A., & McEver, R.P. (1992). Identification of a specific glycoprotein ligand for P-selectin (CD62) on myeloid cells. J. Cell Biol. 118, 445-456.

Mulligan, M.S. & Ward, P.A. (1993). Role of selectins in acute inflammatory lung injury in rats. FASEB J. 7, A3707.

Mulligan, M.S., Varani, J., Dame, M.K., Lane, C.L., Smith, C.W., Anderson, D.C., & Ward, P.A. (1991). Role of endothelial-leukocyte adhesion molecule 1 (ELAM-1) in neutrophil-mediated lung injury in rats. J. Clin. Invest. 88, 1398-1406.

Mulligan, M.S., Watson, S.R., Fennie, C., & Ward, P.A. (1993). Protective effects of selectin chimeras in neutrophil-mediated lung injury. J. Immunol. 151, 6410-6417.

Munro, J.M., Pober, J.S., & Cotran, R.S. (1991). Recruitment of neutrophils in the local endotoxin response: Association with *de novo* endothelial expression of endothelial leukocyte adhesion molecule-1. Lab. Invest. 64, 295-299.

Newman, W., Beall, L.D., Carson, C.W., Hunder, G.G., Graben, N., Randhawa, Z.I., Gopal, T.V., Wiener-Kronish, J., & Matthay, M.A. (1993). Soluble E-selectin is found in supernatants of activated endothelial cells and is elevated in the serum of patients with septic shock. J. Immunol. 150, 644-654.

Norgard, K.E., Moore, K.L., Diaz, S., Stultz, N.L., Ushiyama, S., McEver, R.P., Cummings, R.D., & Varki, A. (1993). Characterization of a specific ligand for P-selectin on myeloid cells. A minor glycoprotein with sialylated O-linked oligosaccharides. J. Biol. Chem. 268, 12764-12774.

Norman, K.E., Moore, K.L., McEver, R.P., & Ley, K. (1995). Leukocyte rolling *in vivo* is mediated by P-selectin glycoprotein ligand-1. Blood 86, 4417-4421.

Ord, D.C., Ernst, T.J., Zhou, L.J., Rambaldi, A., Spertini, O., Griffin, J.D., & Tedder, T.F. (1990). Structure of the gene encoding the human leukocyte adhesion molecule-1 (TQ-1, Leu-8) of lymphocytes and neutrophils. J. Immunol. 265, 7760.

Palecanda, A., Walcheck, B., Bishop, D.K., & Jutila, M.A. (1992). Rapid activation independent shedding of LECAM-1 induced by crosslinking of the surface antigen. Eur. J. Immunol. 22, 1279-1286.

Phillips, M.L., Nudelman, E., Gaeta, F., Perez, M., Singhal, A.K., Hakamori, S., & Paulson, J.C. (1990). ELAM-1 mediates cell adhesion by recognition of a carbohydrate ligand, sialyl-Lex. Science (Wash. D.C.) 250, 1130-1132.

Picker, L.J., Michie, S.A., Rott, L.S., & Butcher, E.C. (1990). A unique phenotype of skin-associated lymphocytes in humans: Preferential expression of the HECA-452 epitope by benign and malignant T cells at cutaneous sites. Am. J. Pathol. 136, 1053-1068.

Picker, L.J., Warnock, R.A., Burns, A.R., Doerschuk, C.M., Berg, E.L., & Butcher, E.C. (1991a). The neutrophil selectin LECAM-1 presents carbohydrate ligands to the vascular selectin ELAM-1 and GMP-140. Cell 66, 921-934.

Picker, L.J., Kishimoto, T.K., Smith, C.W., Warnock, R.A., & Butcher, E.C. (1991b). ELAM-1 is an adhesion molecule for skin-homing T cells. Nature (London) 349, 796-799.

Picker, L.J. & Butcher, E.C. (1992). Physiological and molecular mechanisms of lymphocyte homing. Ann. Rev. Immunol. 10. 561-591.

Picker, L.J., Treer, J.R., Ferguson-Darnell, B., Collins, P.A., Buck, D., & Tertappen, L.W.M.M. (1993a). Control of lymphocyte recirculation in Man. I. Differential regulation of the peripehral lymph node homing receptor L-selectin on T cells during the virgin to memory cell transition. J. Immunol. 150, 1105-1121.

Picker, L.J., Treer, J.R., Ferguson-Darnell, B., Collins, P.A., Bergstresser, P.R., & Terstappen, L.W.M.M. (1993b). Control of lymphocyte recirculation in man. II Differential reguation of the cutaneous lymphocyte-associated antigen, a tissue-selective homing receptor for skin homing T cells. J. Immunol. 150, 1122-1136.

Pober, J.S. (1988). Cytokine-mediated activation of vascular endothelium: Physiology and pathology. Am. J. Pathol. 133, 426.

Pober, J.S., & Cotran, R.S. (1991). What can be learned from the expression of endothelial adhesion molecules in tissues? Lab. Invest. 64, 301-305.

Postigo, A.A., Garcia-Vicuna, R., Diaz-Gonzalez, F., Arroyo, A.G., De Landazuri, M.O. Chi-Rosso, G., Lobb, R.R., Laffon, A., and Sanchez-Madrid, F. (1992). Increased binding of synovial T lymphocytes from rheumatoid arthritis to endothelial-leukocyte adhesion molecule-1 (ELAM-1) and vascular cell adhesion molecule-1 (VCAM-1). J. Clin. Invest. 89, 1445-1452.

Rosen, S.D. (1989). Lymphocyte homing: Progress and prospects. Curr. Opin. Cell. Biol. 1, 913-919.

Rosen, S.D. (1990). The LEC-CAMs: An emerging family of cell adhesion receptors based upon carbohydrate recognition. Am. J. Respir. Cell. Mol. Biol. 3, 397-402.

Sako, D., Chang, X-J., Barone, K.M., Vachino, G., White, H.M., Shaw, G., Veldman, G.M., Bean, K.M., Ahern, T.J., Furie, B., Cumming, D.A., & Larsen, G.R. (1993). Expression cloning of a functional glycoprotein ligand for P-selectin. Cell 75, 1179-1186.

Sanders, W.E., Raymond, W.W., Ballantyne, C.M., & Beaudet, A.L. (1992). Molecular cloning and analysis of in vivo expression of murine P-selectin. Blood 80, 795-800.

Schleiffenbaum, B., Spertini, O., & Tedder, T.F. (1992). Soluble L-selectin is present in human plasma at high levels and retains functional activity. J. Cell. Biol. 119, 229-238.

Seekamp, A., Till, G.O., & Ward, P.A. (1993). Role of selectins in ischemia/reperfusion injury. FASEB J. 7, A2858.

Shimizu, Y., Shaw, S., Graber, N., Gopal, T.V., Horgan, K.J., Van Seventer, G.A., & Newman, W. (1991). Activation-independent binding of human memory T cells to adhesion molecule ELAM-1. Nature (London) 349, 799-803.

Shimizu, Y., & Shaw, S. (1993). Mucins in the mainstream. Nature 366, 630-631.

Siegelman, M.H., van de Rijn, M., & Wiessman, I.L. (1989). Mouse lymph node homing receptor cDNA encodes a glycoprotein revealing tandem interaction domains. Science (Wash. D.C.) 243, 1165-1172.

Siegelman, M.H., Cheng, I.C., Weissman, I.L., & Wakeland, E.K. (1990). The mouse lymph node homing receptor is identical with the lymphocyte cell surface marker Ly-22: Role of the EGF domain in endothelial binding. Cell 61, 611.

Smith, C.W., Kishimoto, T.K., Abbassi, O., Hughes, B., Rothlein, R., McIntire, L.V., Butcher, E.C., & Anderson, D.C. (1991). Chemotactic factors regulate lectin adhesion molecule 1 (LECAM-1)-dependent neutrophil adhesion to cytokine-stimulated endothelial cells in vitro. J. Clin. Invest. 87, 609-618.

Spertini, O., Luscinskas, F.W., Kansas, G.S., Munro, J.M., Griffin, J.D., Gimbrone, M.A., Jr., & Tedder, T.F. (1991a). Leukocyte adhesion molecule-1 (LAM-1, L-selectin) interacts with an inducible endothelial cell ligand to support leukocyte adhesion. J. Immunol. 147, 2565-2573.

Spertini, O., Kansas, G.S., Reimann, K.A., MacKay, C.R., & Tedder, T.F. (1991b). Function and evolutionary conservation of distinct epitopes on the leukocyte adhesion molecule-1 (TQ-1 Leu-8) that regulates leukocyte migration. J. Immunol. 147, 942-953.

Springer, T.A. (1990). Adhesion receptors of the immune system. Nature (London) 346, 425-434.

Springer, T.A., & Lasky, L.A. (1991). Sticky sugars for selectins. Nature (London) 349, 196-197.

Stamper, H.B., & Woodruff, J.J. (1976). Lymphocyte homing into lymph nodes: in vitro demonstration of the selective affinity of recirculating lymphocytes for high endothelial venules. J. Exp. Med. 144, 828-833.

Steegmaier, M., Levinovitz, A., Isenmann, S., Borges, E., Lenter, M., Kocher, H.P., Kleuser, B., & Vestweber, D. (1995). The E-selectin-ligand ESL-1 is a variant of a receptor for fibroblast growth factor. Nature 373, 615.

Sternberg, P.E., McEver, R.P., Shuman, M.A., Jacques, Y.V., & Bainton, D.F. (1985). A platelet alpha-granule membrane protein (GMP-140) is expressed on the plasma membrane after activation. J. Cell Biol. 101, 880-886.

Streeter, P.R., Rouse, B.T.N., & Butcher, E.C. (1988). Immunohistologic and functional characterization of a vascular addressin involved in lymphocyte homing into peripheral lymph nodes. J. Cell. Biol. 107, 1853-1862.

Stoolman, L., Grober, J., Bowen, B., Reddy, P., Shih, J., Thompson, C., Fox, D.A., & Ebling, H. (1992). Selectin-dependent monocyte adhesion in frozen sections of Rheumatoid synovitis. In: *Structure, Function, and Regulation of Molecules Involved in Leukocyte Adhesion* (Lipsky, P.E., & Rothlein, R., Kishimoto, T.K., Faanes, R.B., & Smith, C.W., Eds.). Springer-Verlag, New York. pp. 220-216.

Tedder, T., Isaacs, C., Ernst, T., Demetri, G., Alder, G., & Disteche, C. (1989). Isolation and chromosomal localization of cDNAs encoding a novel human lymphocyte cell surface molecule, LAM-1: Homology with the mouse lymphocyte homing receptor and other human adhesion molecules. J. Exp. Med. 170, 123-133.

von Andrian, U.H., Chambers, J.D., McEvoy, L., Bargatze, R.F., Arfors, K.-E., & Butcher, E.C. (1991). Two step model of leukocyte-endothelial cell interaction in inflammation: Distinct roles for LECAM-1 and leukocyte beta-2 integrins in vivo. Proc. Natl. Acad. Sci. USA 88, 1538-1542.

von Andrian, U.H., Hansell, P., Chambers, J.D., Berger, E.M., Filho, I.T., Butcher, E.C., & Arfors, K.-E. (1992). L-selectin function is required for beta 2-integrin-mediated neutrophil adhesion at physiological shear rates in vivo. Am. J. Physiol. 263, 1034-1044.

Walcheck, B., White, M., Kurk, S., & Jutila, M.A. (1992a). Characterization of the bovine peripheral lymph node homing receptor: A lectin cell adhesion molecule. European J. Immunol. 22, 469-476.

Walcheck, B., & Jutila, M.A. (1994). Gamma/delta T cells express high levels of functional peripheral lymph node homing receptor. International Immunol. 6, 81-91.

Walcheck, B., Watts, G., & Jutila, M.A. (1993). Bovine gamma/delta T cells bind E-selectin via a novel glycoprotein receptor. First characterization of a lymphocyte-E-selectin interaction in an animal model. J. Exp. Med. 178, 853-863.

Walz, G., Aruffo, A., Kolanus, W., Bevilacqua, M., & Seed, B. (1990). Recognition by ELAM-1 of the SSialyl-Lex determinant on myeloid and tumor cells. Science (Wash D.C.) 250, 1132-1135.

Watson, S.R., Fennie, C., & Lasky, L.A. (1991a). Neutrophil influx into an inflammatory site inhibited by a soluble homing receptor-IgG chimera. Nature (London) 349, 164-167.

Watson, S.R., Imai, Y., Fennie, C., Geoffrey, J., Singer, M., Rosen, S.D., & Lasky, L.A. (1991b). The complement binding-like domains of the murine homing receptor facilitate lectin activity. J. Cell. Biol. 115, 235-344.

Weller, A., Isenmann, S., & Vestweber, D. (1992). Cloning of the mouse endothelial selectins: Expression of both E- and P-selectin is inducible by tumor necrosis factor. J. Biol. Chem. 287, 15176-15183.

Whelen, J., Ghersa, P., Hooft van Huijsduijnen, R., Gray, J., Chandra, G., Talabot, F., & Delamater, J.F. (1991). A NK kappa B-like factor is essential but not sufficient for cytokine induction of endothelial leukocyte adhesion molecule 1 (ELAM-1) gene transcription. Nucleic Acid Res. 19, 2645-2653.

Wolber, F., Craig, R., Abbassi, O., Ballew, J., Lobb, R., & Stoolman, L. (1993). VLA-4 mediates lymphocyte binding to endothelium under shear. FASEB 7, A3704.

DISCOVERY AND ANALYSIS OF THE CLASSICAL CADHERINS

Gerald B. Grunwald

Advances in Molecular and Cell Biology,
Volume 16, pages 63-112.

ISBN: 0-7623-0143-0.

It would appear that the principles of cell-specific movements, selective adhesions and cavitation ... help to elucidate the factors instrumental in the normal formation of germ layers and their further segregation.... At present, it would be futile to speculate further upon the possible subcellular factors that are engaged in cellular adhesiveness. It should be pointed out however that this principle is of universal significance in morphogenesis, and that, in connection

> *with directed cell movements, it is deserving of more attention than it has received.*
>
> —P.L. Townes and J. Holtfreter (1955)

I. INTRODUCTION

Johannes Holtfreter's pioneering experiments in embryology clearly identified differential cell adhesion as a morphogenetic mechanism. While the technology of the time precluded identification of the molecules underlying this phenomenon, if he could have invented a family of proteins to carry out these processes, the molecules would be the cadherins. The cadherins are cell surface integral membrane glycoproteins, found in the wide variety of organisms throughout the animal kingdom, that mediate calcium-dependent adhesive interactions critical for the stable associations of cells in solid tissues. The numerous members of the cadherin gene superfamily loom large at all stages of the life cycle of multicellular organisms. From compaction of the morula, the first discernible morphogenetic event in mammalian embryogenesis, to the establishment of discrete germ layers, tissues, and organs during subsequent development, the cadherins play a prominent role in orchestrating the cellular rearrangements that form the basis of morphogenesis. Cadherins also have important functions in the maintenance of mature tissues throughout the life of an individual, and their malfunction or misexpression may contribute to unfortunate pathogenetic changes that may herald disease and premature demise. This review will focus on the classical cadherins, those members of the cadherin gene superfamily which include the first described and most structurally conserved of these proteins and which to date have been found only in vertebrate animals. However, as discussed in this and other companion chapters in this volume, the classical cadherins are part of a growing cadherin gene superfamily comprising a phylogenetically ancient, large and diverse group of molecules with a rich variety of structural motifs, spatiotemporal expression patterns and functional roles in the development and mature function of vertebrate and invertebrate organisms alike. Additional perspectives on the cadherins are available from several recent reviews (Cunningham and Edelman, 1990; Geiger et al., 1990; Geiger and Ayalon, 1992; Grunwald, 1991, 1992, 1993; Kemler et al., 1990; Kemler, 1992; Ranscht, 1991; Takeichi, 1988, 1990, 1991).

II. IDENTIFICATION OF THE CADHERINS AS MEDIATORS OF CALCIUM-DEPENDENT MORPHOGENETIC CELL ADHESION

A. The Search for Cell Adhesion Molecules

Cell adhesion research has a long and rich history, and the conceptual and experimental background to a century of vertebrate embryonic cell adhesion research that led to the identification of cadherin and other intercellular adhesion molecules has recently been reviewed (Grunwald, 1991). Holtfreter's classic experiments with recombined tissue fragments and subsequently with mixtures of dissociated cells, both derived from amphibian early embryonic tissues, clearly established the existence of cell-autonomous surface-mediated affinities which resulted in their assuming non-random associations and sorting out into appropriate tissue-like patterns (Holtfreter, 1939; Townes and Holtfreter, 1955). Later experiments using trypsin-dissociated tissues from more developmentally advanced avian and mammalian embryos demonstrated that cells from even well-differentiated tissues possessed similar selective adhesive properties (Moscona, 1961, 1962). Many early experiments had suggested that calcium was essential for the maintenance of tight cell associations, and this was also observed to be the case in these cell aggregation studies. Additionally, it had become the practice to carry out trypsin dissociation of tissues in the absence of calcium, since this resulted in a more efficient preparation of single cells. The inclusion of calcium during trypsinization resulted in the production of a population of cells that retained a greater degree of residual adhesiveness (Steinberg et al., 1973). The biochemical basis for this was first demonstrated by Takeichi (1977), who found that the presence of calcium during tryptic digestion of cells and tissues prevented the removal of specific cell surface proteins, apparently by affecting the protein conformation and subsequent tryptic sensitivity. Retention of these proteins on the cell surface correlated with the ability of the cells to adhere in a calcium-dependent manner. In the absence of calcium, even low levels of trypsin inactivated this adhesive system, while leaving intact a functionally distinct calcium-independent cell-cell adhesive system. The expression of such dual adhesive systems among cells was described by several laboratories (Urushihara et al., 1979; Grunwald et al., 1980; Brackenbury et al.,

1981; Magnani et al., 1981). Of these two adhesive systems, it was demonstrated that the calcium-dependent system seemed to play the dominant role in forming tighter adhesions which were essential for the formation of histotypic cell interactions, and that this adhesive system was subject to developmental regulation (Grunwald et al., 1981). It is now recognized that the calcium-dependent interactions are mediated by the cadherins, while the calcium-independent interactions are mediated in large part by adhesion proteins which are members of the immunoglobulin gene superfamily, discussed elsewhere in this volume.

B. Identification of Cadherins by Biochemical, Immunological, and Functional Assays

The first cadherins to be identified were found through a variety of approaches in a number of different experimental systems. The protection of cadherins by calcium against proteolytic digestion was used to advantage in the identification of the original cadherins. Cell surface proteins whose presence correlated with the expresion of the calcium-dependent adhesive system were identified by direct biochemical comparisons of cell surface protein patterns on adhesive and non-adhesive cells that had been trypsinized in the presence and absence of calcium, respectively. In conjunction with many of these studies, the use of adhesion function-blocking antisera and later monoclonal antibodies to these proteins further implicated them as the molecules responsible for mediating calcium-dependent cell adhesive interactions, and they were dubbed "cadherins" (Yoshida-Noro et al., 1984). Among the earliest cadherins identified by this combined functional, biochemical, and immunochemical approach was E-cadherin, so named for its prevalent expression in epithelial cells. It was identified both in teratocarcinoma cells as E-cadherin (Yoshida and Takeichi, 1982; Yoshida-Noro et al., 1984) and independently identified and named uvomorulin due to its demonstrated role in mammalian embryo compaction (Kemler et al., 1977; Hyafil et al., 1980; Peyrieras et al., 1983). Human E-cadherin was named cell-CAM 120/80 upon its identification in mammary carcinoma cells (Damsky et al., 1983) and as Arc-1 in MDCK cells (Behrens et al., 1985). The presumed avian form of E-cadherin, refered to as L-CAM, was named following its identification in chick embryo liver cells (Bertelloti et al., 1980; Gallin et al., 1983). N-cadherin was also identified by combined functional, biochemical and immunological criteria, and was originally described in the embryonic chick neural retina and refered to as gp130/4.8 (Grunwald et al., 1982; Cook and Lilien, 1982; Lagunowich and Grunwald, 1989) and later as NcalCAM (Crittenden et al., 1988). N-cadherin was so named upon its identification with a monoclonal antibody and due to its prevalent expression in neural tissues (Hatta et al., 1985; Hatta and Takeichi, 1986). In the interim, it had also been identified as A-CAM, an adherens junction protein from chick heart muscle cell intercalated discs (Volk and Geiger, 1984, 1986). Monoclonal antibodies which blocked adhesion of extraembryonic cells led to the identification of P-cadherin, found in the placenta (Nose and Takeichi, 1986). Immunological and peptide mapping studies led to the identification of a close analogue of N-cadherin called CRM-L for cadherin related molecule in liver (Crittenden et al., 1988).

C. Identification of Cadherins by Molecular Genetic Techniques

Following the identification of cadherin proteins, partial amino acid sequence data and immunological screening of cDNA expression libraries led to the cloning of cDNAs for L-CAM (Gallin et al., 1987), E-cadherin (Schuh et al., 1986; Nagafuchi et al., 1987), P-cadherin (Nose et al., 1987), and N-cadherin (Hatta et al., 1988). This permitted the determination of basic cadherin protein structures and clearly indicated that these various proteins were closely related members of a multigene family. This high homology among cadherins also provided for the development of probes for cDNA library screening and the identification of both known cadherins in additional species and of novel cadherins. The chick N-cadherin cDNA was used as a probe to clone the murine N-cadherin cDNA (Miyatani et al., 1989) and the *xenopus* EP-cadherin (E- and P-cadherin-like) cDNA (Ginsberg et al., 1991). Similarly, a nucleotide probe corresponding to the cytoplasmic domain of mouse E-cadherin was used to screen chick embryo retina cDNA libraries and identified clones for R-cadherin (Inuzuka et al., 1991a). A PCR-generated human N-cadherin probe was used to clone the full-length human N-cadherin cDNA (Reid and Hemperly, 1991). Library screening with antibodies prepared against a detergent-insoluble, concanavalin A-binding glycoprotein from chick brain led to the cloning of the unique truncated T-cadherin (Ranscht and Dours-Zimmerman,

1991). An antiserum raised against the synthetic peptide corresponding to the cytoplasmic domain of L-CAM was used to screen an embryonic chick brain cDNA library and led to the identification of B-cadherin (Napolitano et al., 1991). Naturally occurring antibodies from patients with pemphigus vulgaris, an autoimmune epidermal blistering disease, were used to screen human keratinocyte cDNA libraries and identify PVA, the pemphigus vulgaris antigen, which was identified as a cadherin (Amagai et al., 1991).

As above for cDNA cloning, each advance in molecular genetics has been in turn applied to cadherins, most recently PCR analysis to search for previously unidentified cadherins. PCR analysis led to the identification of M-cadherin, expressed in muscle (Donalies et al., 1991), and of bovine N- and P-cadherin in endothelial cells (Liaw et al., 1990). The most striking application of this approach has been the identification of 11 putative cadherins in adult rat brain, three of which corresponded to the previously identified N, E, and P-cadherins, but eight of which appear to be novel (Suzuki et al., 1991). Northern blot analysis indicated that these putative cadherins were indeed expressed in the brain. PCR analysis also led to the identification of the *Drosophila* fat gene product as a cadherin-like protein (Mahoney et al., 1991). This large transmembrane protein, which represents the first identified invertebrate cadherin, contains 34 cadherin domains as well as other structural motifs such as EGF repeats. The previously known gene was named fat because mutations at this locus lead to overgrowth of embryonic tissues, suggesting the protein may have tumor-suppressor function. Structural analyses of independently identified proteins and cDNAs have also identified members of the cadherin family through sequence homologies, as in the case of the desmosomal components desmoglein and desmocollin (Buxton and Magee, 1992; Legan et al., 1992). These cadherin gene superfamily members, while about 30% homologous to N-cadherin at the amino acid level, differ significantly in structure from the classical cadherins, and they are discussed in detail in other chapters of this volume. Recently, scanning of cDNA sequence databanks indicated that the human protooncogene c-ret is structurally related to the cadherins (Schneider, 1992). While it is not clear if c-ret functions as a cadherin, it is of interest because the cytoplasmic domain contains a tyrosine kinase domain, not found in the other known cadherins.

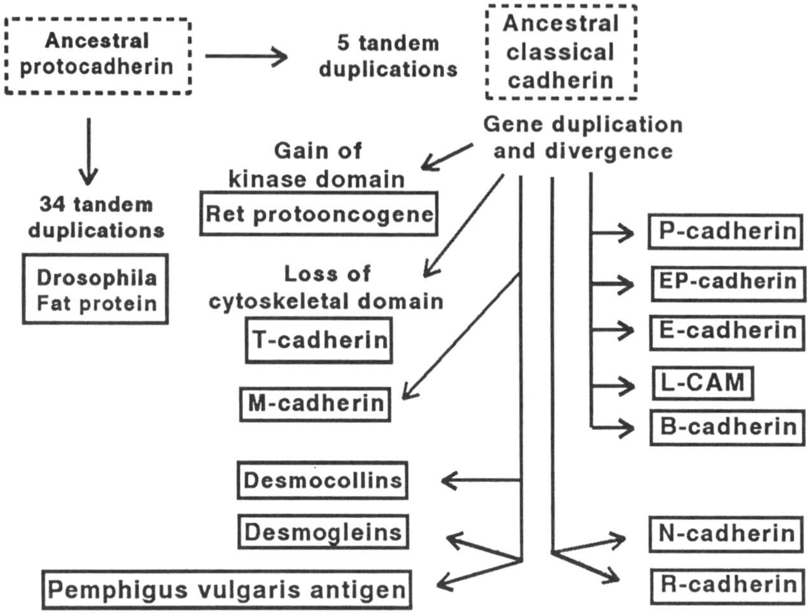

Figure 1. Hypothetical cadherin family tree that relates the classical cadherins (on the right) and more remotely related cadherins which are discussed in the text. They are arranged on the basis of deduced amino acid sequences, based loosely on Pouliot (1992). Evolution of the cadherins appears to have proceeded by a series of gene duplications and recombinations to yield the structurally and functionally diverse cadherin gene superfamily. Reprinted with permission from Grunwald (1993).

D. Cadherin Diversity and Phylogeny

As discussed previously, the original family of the three classical E-, N-, and P-cadherins has now expanded to include several more classical cadherins and a number of interesting variants on the classical motif. In addition to these, enough distant cadherin relatives have been identified to now permit reference to the cadherin gene superfamily. Computer analysis of sequence homologies has led to the construction of a cadherin family tree (Pouliot, 1992). This analysis sorts the cadherins into three groups, consisting of (1) N- and R-cadherins; (2) P-, E-, EP-, and B-cadherins; and (3) the distinct M- and T-cadherins. Evolution of the cadherins seems to have proceeded through a series of gene duplications, perhaps preceded by tandem duplications of an ancestral single extracellular domain containing both a single calcium binding and adhesive recognition site. Such an ancestral protein, comparable to Thy-1 of the immunoglobulin gene superfamily, has yet to be identified, if it still exists. A list of the presently known classical cadherins, and their more distant relations, is presented in Figure 1.

III. ANALYSIS OF CADHERIN STRUCTURE AND BIOSYNTHESIS

A. Basic Organization of Cadherin Proteins

Based mainly upon comparisons of the structures of the original three, N-, E-, and P-cadherins and their homologues, as derived from cDNA-deduced amino acid sequences described in the above references, the classical cadherins are single polypeptide transmembrane glycoproteins with a narrow range of molecular weights averaging 130 kDa and composed of 723-748 amino acids. Each cadherin consists of a large extracellular segment and smaller transmembrane and cytoplasmic segments (Figure 2). The extracellular segment is composed of five distinct repeated domains of about 110 amino acids called EC1 to EC5, starting with the N-terminus. The first four EC domains are highly homologous to one another, while the fifth is the least conserved. Among the extracellular domains, EC1 is the most highly conserved between different cadherins and contains the critical functional site for adhesive recognition. This domain is also the target of most, but not all, adhesion function-blocking antibodies. However, the cytoplasmic domain was found to be the most highly conserved region of the cadherins overall, which immediately suggested an important role integrating extracellular and intracellular cadherin functions. The functions of these various cadherin regions are discussed in detail ahead.

B. Cadherin Biosynthesis

Proteolysis of cadherin precursors at the amino terminus prior to their mobilization to the cell surface appears to be a typical component of cadherin biosynthesis. As studied in detail for E-cadherin, this results in conversion of a 140 kDa precursor into the

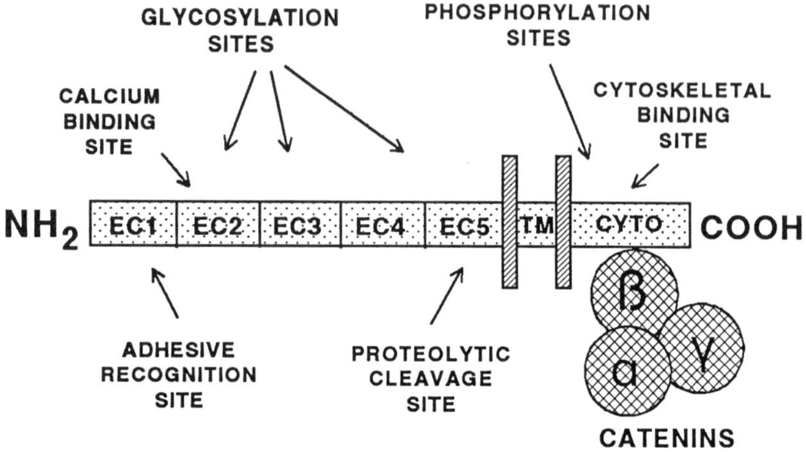

Figure 2. Structural organization of the classical cadherins. Indicated are the five extracellular (EC1-EC5), transmembrane (TM), and cytoplasmic (CYTO) domains. The extracellular domains include multiple glycosylation sites, a proteolytic cleavage site near the cell membrane, and functionally critical calcium binding and adhesion recognition sites. The cytoplasmic domain includes sites critical for interaction with the catenin cytoskeletal proteins. The cadherins and catenins both contain phosphorylation sites which regulate their function in intercellular junctions. See text for details. Reprinted with permission from Grunwald (1993).

mature 120 kDa protein (Ozawa and Kemler, 1990). In this study, site-directed mutagenesis indicated that failure of this cleavage to occur results in the production of inactive cadherins that are mobilized to the cell surface but are unable to mediate adhesions. These cadherins could be activated by removal of the prepeptide region by exogenous proteases. As opposed to the mature protein, precursor regions of the cadherins have little homology between cadherin subtypes.

Biosynthesis and turnover has been studied for both E- and N-cadherin, which show similar profiles. Pulse-chase analysis of E-cadherin in MDCK cells indicated a peak accumulation time of one to two hours and a half life of about five hours (Shore and Nelson, 1991). Similar studies of N-cadherin in the embryonic chick retina also indicated it takes about two hours for the peak accumulation of labelled N-cadherin, which then disappears with a half-life of about five hours (Paradies and Grunwald, 1993). As discussed below, N-cadherin turnover appears to be mediated by a cell surface metalloproteolytic activity that results in the release of a soluble 90 kDa N-terminal fragment called NCAD90 which retains biological activity (Roark et al., 1992; Paradies and Grunwald, 1993). It has been suggested from experiments on E-cadherin synthesis, transport, and turnover that differential turnover of E-cadherin at different regions of cells may contribute to the development of epithelial cell polarity, since E-cadherin delivered to the apical membrane was rapidly removed, while E-cadherin delivered to the basolateral cell membrane had a longer half-life (Wollner et al., 1992).

C. Calcium Binding Sites

As discussed previously, the ability of cadherins to bind calcium in the extracellular domain, resulting in conformational changes that affect adhesive function and protease resistance, were fundamental in the initial identification of cadherins. Putative calcium binding sites rich in asparagine residues were identified in the extracellular domain (Ringwald et al., 1987). The function of these sites has been explored through the use of synthetic peptides and site-directed mutagenesis studies (Ozawa et al., 1990a). Synthetic peptides corresponding to a single calcium-binding domain were shown to bind calcium and undergo conformation changes upon binding. This binding was shown to be dependent on a critical aspartic acid residue. Furthermore, when this residue, at site 134 in the EC2 domain of intact E-cadherin, was changed to lysine, the expressed cadherin was found to be much more sensitive to proteolysis and inactive in formation of adhesive bonds. This calcium-binding site is highly conserved in all cadherins (Table 1).

D. Adhesive Recognition Sites

During the several decades of cell adhesion research preceding the identification of discrete adhesion molecules, a major debate ensued over the relative importance of qualitative factors (Moscona, 1962) versus quantitative factors (Steinberg, 1963) in the determination of adhesive preferences among cells (reviewed in Grunwald, 1991). Analyses of cadherin function have shed light on this issue, and these studies indicate that both parameters may influence cell adhesive preferences.

Table 1. Amino Acid Homologies Among Cadherins

Cadherin Type	*Species Source*	*Recognition Site*			*Calcium Site*		
N	CHICK	RE	HAV	DV	TVTAI	D	ADDP
N	HUMAN	GA	HAV	DI	TVTAI	D	ADDP
N	MOUSE	RA	HAV	DI	TVTAI	D	ADDP
N	BOVINE	RA	HAV	DI	TVTAI	D	ADDP
N	XENOPUS	RA	HAV	DV	TVTAI	D	GDDP
R	CHICK	RA	HAV	DM	TVTAN	D	ADDS
E	MOUSE	YS	HAV	SS	KVSAT	D	ADDD
E	HUMAN	FS	HAV	SS			
B	CHICK	YS	HAV	SE	RVNAT	D	ADDD
L-CAM	CHICK	LS	HAV	SA	TVNAT	D	ADDA
P	MOUSE	YG	HAV	SE	QVTAT	D	EDDA
P	HUMAN	FG	HAV	SE	QVTAT	D	EDDA
EP	XENOPUS	SS	HAV	SE	AVSAT	D	EDDA
T	CHICK	EV	EVT	DL	RMTAF	D	ADDP
5	HUMAN	DK	TGE	EN	SVTAV	D	ADDP
M	MOUSE	RA	FAL	DL	RAEAT	D	ADDP
DG1	HUMAN	YC	RAL	NS	ILNAT	D	ADEP
DG1	BOVINE	YO	RAL	NS	VLNAT	D	ADEP
DG2,3	HUMAN	IA	FAT	TP	QVCAT	D	KDEP
DG2,3	BOVINE	YA	YAT	TA	KVTAI	D	LDEP
FAT	DROSOPHILA		?		V A	D	D

Note: Critical amino acids for adhesive recognition and calcium binding are highlighted. All structures are based upon deduced amino acid sequences.

The differential expression of cadherin subtypes has clearly been demonstrated to affect such adhesive choices. The first direct evidence that cadherins function as adhesion molecules was demonstrated using cadherin cDNA-transfected fibroblastic cells which in the absence of exogenously expressed cadherins are not mutually adhesive in a calcium-dependent fashion. Cells transfected to express cadherins became mutually adhesive, were calcium-dependent, and this could be inhibited with the corresponding cadherin type-specific antibody (Nagafuchi et al., 1987; Edelman et al., 1987; Mege et al., 1988; Miyatani et al., 1989). Such cells transfected to express either murine N-, E-, or P-cadherin showed preferential homophilic adhesion, while cells expressing murine N-cadherin or chicken N-cadherin formed mixed aggregates (Miyatani et al., 1989), further indicating the specific homophilic nature of cadherin interaction. However, some potential for heterophilic interaction has been demonstrated between N-cadherin and L-CAM (Volk et al., 1987).

In these experiments, in which lens cells expressing N-cadherin and liver cells expressing L-CAM were mixed, the composite aggregates that formed showed a preferential segregation into the expected homophilic contacts. However, antibody staining demonstrated the presence of some heterophilic contacts containing both cadherin types. More recently, it has been demonstrated that significant heterophilic adhesive affinities exist between N- and R-cadherin, although among these two cadherins homophilic interactions are still predominant (Inuzuka et al., 1991a). As N- and R-cadherin are significantly more homologous than N-cadherin and L-CAM, this indicates that heterophilic cadherin affinities may vary over a continuum that depends in part upon the sequence similarities in the adhesive binding domain and other regions.

Most antibodies that inhibit cadherin function bind near the N-terminus. The specific regions of cadherins that mediate adhesive recognition, and their amino acid sequences, have been identified by molecular genetic modifications that either delete specific cadherin domains or result in the exchange of domains between different cadherins. Chimeric cadherins representing various combinations of E- and P-cadherin domains, and expressed in transfected cells, demonstrated that the cadherin-type selectivity was largely determined by sequences within the 113-amino acid N-terminal domain (Nose et al., 1990). Swapping of domains for E- and P-cadherin resulted in the generation of chimeric molecules whose adhesive selectivity was dependent on the extracellular EC1 domain. Point mutations further located the key regions for adhesive binding specificity between amino acid residues 78-83, and the targets of adhesion-inhibiting antibodies at amino acids 16 or 31, depending on the antibody and cadherin subtype. The amino acid sequence histidine-alanine-valine (HAV) was found to be conserved in the classical cadherins within the binding specificity region, and site directed mutagenesis which altered the immediately flanking amino acid residues resulted in alterations in cadherin subclass recognition (Table 1). The HAV-containing regions of cadherins have homologies to members of the fibroblast growth factor receptor family and influenza hemagglutinins, suggesting that these motifs may be involved in protein-protein interactions among a wide variety of molecules (Blaschuk et al., 1990; Byers et al., 1992). Further supporting the importance of these sites, peptides containing the HAV domain have been shown to inhibit cadherin-mediated processes such as mouse embryo compaction as well as neurite growth on glial substrates (Blaschuk et al., 1990). While these distal cadherin regions are clearly involved in adhesive recognition and binding, allosteric effects also appear to be involved. Structural alterations in other regions of the molecules, such as the extracellular domain just outside the membrane (Ozawa et al., 1990b), or the cytoplasmic domain (see following), also affect cadherin binding activity.

The above studies clearly indicate that qualitative differences in adhesion molecule expression can mediate selective adhesion and cell sorting. However, it has also been demonstrated that quantitative differences in cadherin expression may also contribute to variations in adhesive interactions and cell sorting. This was demonstrated by cDNA transfection of fibroblastic cells, and the resulting cell lines, with a range of cadherin expression, were found to sort out within mixed aggregates (Friedlander et al., 1989).

IV. MODULATION OF CADHERIN STRUCTURE AND FUNCTION BY POST-TRANSLATIONAL MODIFICATIONS

A. Glycosylation

Cadherins have been known to be glycoproteins since their initial discovery (Cook and Lilien, 1982; Damsky et al., 1983, Vestweber and Kemler, 1984), but the functional significance of these carbohydrate moieties remains unknown. While sequence analysis of cadherins has indicated the general presence of multiple potential glycosylation sites, structural details are available for only a few cadherins. For example, both E- and N-cadherin have been shown to possess at least four N-asparagine linked oligosaccharides in the extracellular domain with differing endoglycosidase sensitivities (Cunningham et al., 1984; Crittenden et al., 1988). However, pharmacological studies suggest that inhibition of glycosylation has no effect on cadherin function or stability (Vestweber and Kemler, 1984; Shirayoshi et al., 1986).

B. Sulfation

Metabolic labeling studies have shown that E-cadherin (Vestweber and Kemler, 1984) and N-cadherin (Lagunowich and Grunwald,

1991) are sulfated, although the functional significance of cadherin sulfation is not known.

C. Phosphorylation

Cadherins have been known for some time to be phosphoproteins. E-cadherin isolated from human mammary carcinoma cells is phosphorylated on serine (Wheelock et al., 1987), while L-CAM from chick liver is phosphorylated on both serine and threonine (Cunningham et al., 1984). More recently phosphorylation has been implicated in functional regulation of cadherins. As discussed previously, E-cadherin had been implicated in the process of mouse embryo compaction, since anti-E-cadherin antibodies were found to inhibit this process. However, E-cadherin was known to be expressed prior to compaction, which indicated that perhaps post-translational modifications could be responsible. Activation of protein kinase C was shown to induce premature compaction and cause a redistribution of E-cadherin to cell-cell junctions, while inhibition of protein kinase C prevented normal compaction (Winkel et al., 1990). The premature compaction was still inhibited by anti-E-cadherin antibodies. This work did not explore whether E-cadherin was a direct target of PKC-induced changes. A subsequent study explored the phosphorylation state of E-cadherin in early mouse embryos, and found that while E-cadherin is only weakly phosphorylated up to the four-cell stage, it becomes phosphorylated at the eight-cell stage just prior to compaction (Sefton et al., 1992). Again, the total level of E-cadherin changed little over this period. These results, which link changes in cell adhesiveness, E-cadherin distribution, and phosphorylation state suggest that post-translational control of E-cadherin may be critical to aspects of its function.

A number of studies utilizing Rous sarcoma virus (RSV) transformed cells, which express the potent tyrosine kinase v-src, have explored the relationship between phosphorylation state of proteins and cell adhesive interactions. Lens cells express high levels of N-cadherin which is localized to adherens junctions. When chick embryo cells were transformed with temperature sensitive RSV, when shifted to the permissive temperature these cells, which are normally epithelial, assumed a fibroblastic appearance, lost adherens junctions, and N-cadherin was redistributed to a diffuse cell-surface organization (Volberg et al., 1991). These studies indicated that protein components of adherens junctions were major locations of phosphotyrosine-containing proteins, although N-cadherin was concluded to not be one of them, since it was not bound to anti-phosphotyrosine immunoaffinity supports. The effect of v-src and v-fos on cadherin-mediated adhesion was also explored among rat fibroblasts, which express endogenous P-cadherin, as well as their transfected counterparts which also expressed E-cadherin (Matsuyoshi et al., 1992). Again, transformation was found to have no effect on overall levels of expression of the cadherins, and no effect was observed on the rate of cell-cell adhesion as measured in suspension assays. However, the morphology of cell aggregates was altered resulting in looser connections between cells. These effects were attributed to changes in tyrosine kinase activity, since it was found that among the transformed cells inhibitors of these enzymes inducted tighter cell associations, while tyrosine phosphatase inhibitors reduced cadherin-mediated adhesion. In this study, cadherin and catenin phosphorylation were examined by immunoblotting with anti-phosphotyrosine antibodies, and it was found that, whereas control cells exhibited no tyrosine phosphate on cadherins, and little on catenins, transformed cells have highly tyrosine-phosphorylated β-catenin and weakly phosphorylated E-cadherin.

A subsequent study of chick embryo fibroblasts, which express N-cadherin, demonstrated that RSV-transformation of these cells did not alter expresion of N-cadherin but did suppress N-cadherin function (Hamaguchi et al., 1993). Transformation also resulted in enhanced tyrosine phosphorylation of both N-cadherin and catenins, but these proteins were still found to exist in a complex, suggesting that the inhibitory effects of phosphorylation on adhesive interactions occur via modulation of other aspects of cadherin-catenin function. Similar results have been reported for v-src-transformed MDCK cells (Behrens et al., 1993).

A direct demonstration of cadherin adhesive function by phosphorylation changes will require specific identification of the modification sites and their alteration through site-directed mutagenesis. However, the above results would predict that kinase inhibitors might serve to stabilize cell adhesions. Indeed, it has been demonstrated in MDCK cells that pharmacologic inhibition of kinase activity prevents the disruption of cell-cell contacts and cadherin redistribution normally induced by calcium depletion from the medium (Citi, 1992). These studies strongly indicate that tyrosine phosphorylation of one or more members of cadherin-catenin complexes may be an important control point for the association and function of these molecules. Studies currently underway in this author's laboratory are directed at addressing these issues in primary embryonic cells as opposed to cell lines which have been virally transformed and express high levels of exogenous tyrosine kinases. N-cadherin was shown to be phosphorylated in retina, brain, heart and lens, that express high levels of this protein (Lagunowich and Grunwald, 1991). In contrast to the above studies on src-transformed cells, this study suggested that a positive corrclation existed between the extent of phosphorylation and the insolubility of N-cadherin in non-ionic detergents and junctional association. More recent studies have directly analyzed the phosphorylation of N-cadherin in these tissues, and indicate that in primary embryonic cells, N-cadherin is constitutively phosphorylated on serine, with peptide digests indicating that least two independent sites (Lee and Grunwald, 1993). Treatments with tyrosine phosphatase inhibitors indicate that N-cadherin is also phosphorylated on tyrosine on a third independent peptide site. This indicates that N-cadherin is a substrate for multiple endogenous kinases, including both serine/threonine and tyrosine kinases, and that the functional state of cadherins may depend on the balance between a number of potential phosphorylation sites.

D. Proteolysis

Several types of proteolytic processing have been described for the cadherins, and they appear to be targets of proteolysis both intracellularly and extracellulary and by exogenous as well as endogenous proteases. Intracellular proteolysis of cadherin precursors to yield the mature protein has already been described. With respect to the extracellular domain, as mentioned previously, removal of calcium renders cadherins sensitive to digestion by exogenously added proteases. The previously cited studies had shown that in the presence of calcium, cadherins may be completely resistant to proteolysis, or may release an approximately 80-90 kDa soluble N-terminal fragment, depending on the cadherin and whether intact tissues or isolated cells are treated. Antibodies to various cadherins obtained from tissue homogenates often detect a 110 kDa membrane-associated fragment on immunoblots. Cadherins yield such a fragment upon cleavage at intracellular sites by calcium-activated proteases in tissue homogenates (Covault et al., 1991), although it is not clear if this type of processing occurs in intact cells or tissues. Cleavage of mature N-cadherin by endogenous proteases has been suggested to be a mechanism for regulation of its expression and function. Cultured lens cells produce a 78 kDa membrane-bound fragment when maintained in low-calcium medium, while a 97 kDa fragment is released into the culture medium even in the presence of calcium (Volk et al., 1990). This study further correlated proteolytic turnover of N-cadherin with the dispersion of epithelial somites in developing chick embryos. Intact chick embryo retinas were known to release a 90 kDa fragment of N-cadherin into medium in culture (Grunwald et al., 1982). The precursor-product relationship of this fragment and 130 kDa N-cadherin was further indicated by turnover studies of iodinated retinal cell surface proteins (Cook et al., 1984). The down-regulation of N-cadherin which occurs during retinal development was shown to itself be protein-synthesis dependent, and inhibitor studies indicated that proteolytic activity was responsible for N-cadherin turnover yielding a soluble 90 kDa N-terminal fragment (Roark et al., 1992). More recent studies have demonstrated through pulse-chase metabolic studies that there is a direct precursor-product relationship between N-cadherin and this NCAD90 protein, that NCAD90 occurs *in vivo,* and that this turnover is mediated by a cell surface-associated metalloprotease (Paradies and Grunwald, 1993; Ferreira et al., 1993). As mentioned earlier, these studies showed that purified NCADD90 retains biological activity and can mediate cell adhesion and neurite growth when attached to a solid substrate (Paradies and Grunwald, 1993). The exact nature of the endogenous proteases responsible for cleavage of cadherins are not known. Recently, however, urokinase plasminogen activator has been shown to be co-distributed with E-cadherin in keratinocytes where it is in a position to influence epithelial cell interactions (Jensen and Wheelock, 1992).

V. INTERACTIONS OF CADHERINS WITH THE CYTOSKELETON AND OTHER PROTEINS

A. Cytoskeletal Interactions and the Catenins

Many of the previously mentioned studies indicated that among the functions of cadherins were an association with cell junctions and the cytoskeleton, since adherens junctions and calcium-dependent adhesiveness were both resistant to proteolysis in the presence of calcium (Grunwald et al., 1981) and cadherins were found to colocalize with adherens junctions and actin bundles (Boller et al., 1985; Hirano et al., 1987; Volk and Geiger, 1984, 1986). The highly conserved nature of the cytoplasmic domain amino acid sequences of cadherins, revealed by the previously mentioned cloning studies, further implicated this region in subserving an important function. Such a linkage was strongly indicated by studies utilizing expression of E-cadherin constructs which were truncated to various degrees in the cytoplasmic domain (Nagafuchi and Takeichi, 1988). These studies indicated that truncated E-cadherin was expressed at the cell surface, but that it could no longer mediate cell-cell binding and that it could now be more efficiently extracted with non-ionic detergents. However, while proteins co-immunoprecipitating with cadherins had been reported previously, their identification and the specific mechanism by which cadherins could directly link to the cytoskeleton remained unclear. Subsequent studies identified three novel proteins of molecular weights 102, 88-92, and 80 kilodaltons, termed α, β and γ catenins, which were found to be conserved in a wide variety of species and whose interaction with E-cadherin was dependent on an intact cytoplasmic domain (Ozawa et al., 1989; Nagafuchi and Takeichi, 1989). α-catenin has been cloned and shown to be related to vinculin and to associate with several different cadherins (Nagafuchi et al., 1991; Herrenknecht et al., 1991). β-catenin has been cloned and shown to be related to both the drosophila protein armadillo and the mammalian junctional protein plakoglobin (McCrea et al., 1991). Armadillo is a segment polarity gene, suggesting that catenins may be involved in developmental pattern formation. However, these three proteins may not be true homologues, but rather may be members of a multigene family (Peifer et al., 1992). The gamma catenin gene has not yet been cloned, but immunological and biochemical studies suggest that it may be more closely related to plakoglobin than β-catenin (Knudsen and Wheelock, 1992). The afore-mentioned studies indicate that catenin-cadherin association appears to be universal, having been demonstrated with numerous cadherins and in different cell types and species, including differentiated tissues such as muscle (Wheelock and Knudsen, 1991).

Biochemical studies have indicated that the stoichiometry of cadherin-catenin association may be one molecule of E-cadherin, one or two molecules of β-catenin, and one molecule of α-catenin (Ozawa and Kemler, 1992). These studies furthermore indicated that β-catenin may be the more direct link with cadherins, and that it becomes associated with E-cadherin even while the latter is still not fully processed. α-catenin becomes associated later, and γ-catenin is the most weakly associated. Some experiments indicate that little if any catenin-free cadherins exist within cells (Ozawa and Kemler, 1992). Multiple cadherins can be simultaneously expressed by cells, and it has been shown that in cultured cells these can be differentially expressed with diffuse pericellular expression as well as in discrete junctional complexes (Salomon et al., 1992). Such different distributions of cadherins may be regulated by catenins, as changes in expression of the latter occur during changes in individual cell polarity as well as during embryonic development (Wheelock, 1990; DeMarais and Moon, 1992).

Transfection of cells with cadherin cDNAs has been shown to up-regulate expression of α-catenin protein, while not affecting mRNA levels, indicating a possible translational regulatory mechanism for maintaining a balanced expression of cadherins and catenins (Nagafuchi et al., 1991). The catenin story is likely to become more complex, as recently a novel predominantly neural subtype of α-catenin has been described (Hirano et al., 1992). In this study a line of lung carcinoma cells, which expresses E-cadherin and β-catenin, but no α-catenin, and normally grows as single cells, was induced to form aggregates when transfected to express neural α-catenin. Thus, the catenins, like the cadherins with which they interact, may also be members of multigene families, raising the potential for complex regulatory cross-interactions among these two protein families.

The importance of cytoplasmic interactions for cadherin function was made clear in two series of experiments where mutant N-cadherin molecules, which lack extracellular domains, were generated and expressed in xenopus embryos (Kintner, 1992) and in a keratinocyte cell line (Fujimori and Takeichi, 1993). In both instances, cell interactions were disrupted, leading to abnormal embryonic development of epithelial tissues *in vivo* and decreased cell interactions *in vitro,* respectively. However, these authors differ in their interpretation of the results, since one set of experiments suggested inhibition mediated by competition for catenins and inhibition of their interaction with endogenous cadherins (Kintner, 1992), while the other experiments suggested no inhibition of catenin interaction, but rather a displacement of endogenous cadherins from their normal junctional sites, at which the mutant cadherin was found, without competition for catenins (Fujimori and Takeichi, 1993). While these different outcomes may simply reflect the use of different experimental systems, these data could also be reconciled if further studies indicate that the results along a continuum of possible outcomes that depend on the relative amounts and subtypes of cadherins and catenins competing within a cell.

Another recent study utilized fibroblastic cells transfected with chimeric cDNAs to express a chimeric receptor containing the cytoplasmic and transmembrane domains of β1-integrin and the extracellular domain of N-cadherin (Geiger et al., 1992). The expressed protein was found to preferentially localize to extracellular matrix adhesion sites, indicating that the cytoplasmic integrin domain was guiding cytoskeletal associations and subcellular distribution. However, transfectants expressing a high level of the chimeric protein assumed a more epithelial morphology and the protein was also found in cell-cell junctions. Interestingly, talin, a cytoskeletal component normally found in cell-matrix junctions, was recruited into these latter junctions. Thus, the localization of adhesion molecules into different types of specialized junctions, and the attendant association with distinct cytoskeletal elements, appears to depend on a combination of factors mediated by interactions with both extracellular and intracellular ligands for such adhesion receptors.

B. Direct Cytoskeletal Interactions of Cadherins Are Not Necessary for All Cadherin Functions

The previously cited experiments in which small peptides containing the HAV domain were found to inhibit cadherin binding indicate that the isolated binding region may retain biological activity. This is also supported by the observation that the soluble extracellular 80 kDa domain of E-cadherin also competitively inhibits cadherin function (Wheelock et al., 1987). These experiments do not address the point of whether positive adhesive interactions can be mediated in the absence of a cytoplasmic domain. Nature has addressed this point in providing an interesting variation on the classical cadherin structural theme in the form of T-cadherin, a truncated cadherin, which contains the typical extracellular cadherin domain organization but lacks both the HAV sequence and the cytoplasmic domain (Ranscht and Dours-Zimmerman, 1991). T-cadherin is linked to the cell surface by a phosphatidylinositol glycan linkage. T-cadherin can mediate homophilic calcium-dependent adhesions among cells in suspension that have been transfected to express it, although it does not accumulate at intercellular boundaries among these cells when they are grown as a monolayer (Vestal and Ranscht, 1992). This study also indicated that T-cadherin appears to be relatively resistant to proteolysis even in the absence of calcium, and is not affected by drugs that perturb the cytoskeleton, which is not surprising since it lacks a cytoplasmic domain.

Additional evidence for cadherin interactions in the absence of a cytoplasmic domain was obtained in experiments where the extracellular domain of N-cadherin, NCAD90, which occurs *in vivo* as a naturally occuring soluble proteolytic turnover product of N-cadherin (Roark et al., 1992), was purified and bound to a solid substrate in adhesion assays (Paradies and Grunwald, 1993). This study demonstrated that the isolated extracellular domain, referred to as NCAD90, could promote both adhesion and neurite growth among retinal neurons. More recently, direct homophilic NCAD90 binding has been demonstrated using covalently modified latex microbeads (Grunwald et al., 1993). Interestingly, as opposed to intercellular cadherin-mediated adhesion, this NCAD90-mediated bead adhesion was found to be calcium-independent, although NCAD90 binds and is protected by calcium against trypsin. This raises the possibility that the calcium dependence of cadherin-mediated cell adhesion may be due in part to intracellular signaling or other functions than that mediated by the extracellular calcium-binding domains, as discussed ahead.

C. Cadherin Interactions with Other Proteins

N-cadherin has been demonstrated to be a substrate for a cell surface N-acetylgalactosaminylphosphotrasferase (GalNAcPTase) whose spatiotemporal expression, like that of N-cadherin, changes during development (Balsamo et al., 1986; Balsamo and Lilien, 1990). This enzyme appears to be structurally coupled to cadherins as the two proteins coimmunoprecipitate. Furthermore, a functional interaction between the GalNAcPTase and cadherins has been demonstrated in several systems. Antibodies to the enzyme uncouple cadherins from the cytoskeleton and inhibit N-cadherin-mediated adhesion among retinal cells (Balsamo et al., 1991). Furthermore, antibodies to the enzyme inhibit E-cadherin-mediated adhesion among pancreatic islet cells (Bauer et al., 1992). While the interaction with N-cadherin appears to be direct as part of a macromolecular complex, antibody perturbation of the enzyme affects not only N-cadherin-mediated neurite growth but that promoted by integrin and immunoglobulin family adhesion molecules as well (Gaya-Gonzalez et al., 1991).

While no direct interactions between cadherins and other cell adhesion molecules have been demonstrated, indirect influences have been suggested by a number of experiments. Conflicting data exists on N-CAM—N-cadherin interaction, since cadherin function could be modulated by N-CAM—mediated interactions among membrane vesicles (Rutishauser et al., 1988), although such interaction was not observed among intact neurons (Doherty et al., 1991b). While the classical cadherins are a major component of adherens junctions, and the related desmogleins and desmocollins are major components of desmosomes, they may act at least indirectly to regulate the formation of other types of intercellular junctions such as gap junctions and tight junctions (Mege et al., 1988; Gumbiner et al., 1988). The connexins, which form gap junctions, are not structurally related to cadherins. However, among cells which express both connexins and N-cadherin, and possess both adherens and gap junctions, antibodies to either protein alone were found to inhibit formation of both types of junctions (Meyer et al., 1992). In addition to the catenins, cadherin-containing protein complexes have been shown to contain ankyrin and fodrin (Nelson et al., 1990). Expression of exogenous E-cadherin in fibroblasts, which as previously mentioned results in their assuming a more epithelial appearance, also results in the redistribution of Na+, K+—ATPase from a diffuse to a polarized membrane distribution, which parallels the normal expression in mesenchymal and epithelial cells, respectively (McNeill et al., 1990). Such a redistribution was not observed in cells transfected to express truncated E-cadherin lacking the cytoskeletal domain. Thus, the cadherins have the potential to affect a wide variety of cellular molecules and overall morphology via their coordination of extracellular and intracellular events.

VI. ORGANIZATION AND REGULATION OF CADHERIN GENES

A. Genomic Organization

Among those cadherins whose gene organization has been explored, results suggest that each cadherin type is encoded by distinct single-copy genes. A possible exception to this appears to be Xenopus N-cadherin, which possesses two pseudo-alleles that are both expressed, although apparently in similar patterns (Simonneau et al., 1992). Detailed information is available on the genomic organization of N-, E-, and P-cadherin, the original three classic cadherins. The structures of the chicken L-CAM (Sorkin et al., 1988) and mouse E-cadherin (Ringwald et al., 1991) genes are highly homologous in terms of exon-intron organization, although the E-cadherin gene is much larger. The E-cadherin gene is located on mouse chromosome 8, and is encoded by 16 exons encompassing over 40 kb of DNA (Ringwald et al., 1991). These studies suggested that the exon structure does not correspond in any obvious way to functional domains of the protein. E-cadherin promoter analysis indicated that tissue-specific epithelial expression is controlled in part within the 5' upstream region which contains motifs for several known transcription factors. Control of epithelial expression has been more closely localized to two regions, one GC-rich basic promoter and a neighboring palindromic enhancer which is structurally and functionally homologous to keratin gene promoter elements (Behrens et al., 1991). More recent analysis of the chick L-CAM gene revealed another cadherin gene upstream named K-CAM, which by sequence analysis is highly homologous to chicken B-cadherin (Sorkin et al., 1991). The mouse P-cadherin gene is encoded by 15 exons which extend over 45 kb of DNA, which includes a very large first intron of 23 kb (Hatta et al., 1991). Strong homologies exist between the P-cadherin gene and that of E-cadherin and L-CAM, and like E-cadherin, P-cadherin is also located on mouse chromosome 8. The M-cadherin gene has also been localized to mouse chromosome 8, which links it to murine E- and P-cadherin, and to human chromosome 16, which also links it to E-cadherin (Kaupmann et al., 1992). The structure of the mouse N-cadherin gene has also recently been analyzed and, unlike E- and P-cadherin, is located on mouse chromosome 18 (Miyatani et al., 1992), while human N-cadherin has been placed on human chromosome 18 (Walsh et al., 1990). Again highly homologous exon-intron organization was found when compared to those of other cadherins. However, the murine N-cadherin gene is huge, encompassing more than 200 kb of DNA which include 16 exons. The large size of the gene is due in part to a large first intron, as for P-cadherin, but in addition there is a uniquely large second intron of more than 100 kb in size. There is also a highly conserved extra 16th exon which appears to have arisen by a recent duplication, but is of unknown function.

B. Transcriptional Regulation of Cadherins

No evidence exists for differential splicing among the classical cadherins, although it has been described for their desmosomal relatives, the desmocollins (Parker et al., 1991). In general, single major mRNA species have been detected on northern blots, which correspond to the single major protein species detected on western blots, and there is no evidence to suggest that the minor additional mRNA species detected are transcribed into identifiable gene products. Nevertheless, the appearance of multiple mRNA bands on Northern blots has been a common observation. For example, the murine N-cadherin mRNA is 4.3 kb in fetal mouse brain and heart, with minor bands at 5.2 and 3.5 kb (Walsh et al., 1990). This study also found N-cadherin mRNA at 5.2, 4.3, and 4.0 kb in C6 glioma cells, at 4.3 and 4.0 kb in human fetal muscle, and 4.3 kb in human fetal brain. Expression of cadherin mRNAs have been examined in the development of a variety of tissues, including retina (Roark et al., 1992), testis (Cyr et al., 1992), hippocampus (Wagner et al., 1992), and whole rat embryos (Chen et al., 1991). These studies indicate both quantitative and qualitative developmental changes in cadherin mRNA expression, and although several minor species of messages were identified in many of these studies, it is not yet clear if this variation is due to species, tissue, or age differences, cross-reactivity with other cadherins, or structural variations such as differential polyadenylation.

VII. EXPRESSION AND FUNCTION OF CADHERINS DURING DEVELOPMENT

A. Cadherins Are Expressed in Complex Patterns During Development

The availability of immunological and molecular genetic probes has permitted extensive analyses of cadherin expression both during development and in mature tissues. These studies have indicated that cadherins are expressed in spatially and temporally dynamic patterns which suggest roles in the histogenesis of a wide variety of embryonic structures, with major changes occurring during active periods of rearrangements of cells resulting in new assemblies and segregations. The largest body of information exists for the three originally identified cadherins, E-, N-, and P-cadherin. Such descriptive studies, of which selected examples follow, have demonstrated that most cadherin names are misnomers, since they are based upon the first tissue in which the respective cadherins were identified. Most cadherins have a complex pattern of expression and can be found in multiple tissues. Also, many tissues express multiple cadherins.

While cadherins are undoubtedly involved in cell signalling events, there is no data to suggest that they play a direct role in early embryonic inductive interactions. Rather, their distinct patterns of expression appear to be a result of such interactions. For example, studies in xenopus show that N-cadherin expression results from the induction of ectoderm prior to neurulation, and no N-cadherin mRNA is detected in early embryos before induction or in isolated ectoderm which does not undergo induction (Detrick et al., 1990). However, in these experiments, N-cadherin could be induced in isolated ectoderm using a heterologous inducer, Hensen's node from the chick embryo. Thus N-cadherin expression appears to follow embryonic induction and precede the morphogenetic changes associated with neural tube formation.

B. E-cadherin Is Expressed Early in Development and Remains Expressed in Epithelia

E-cadherin and its homologues are the first cadherins expressed in avian and mammalian embryonic tissues during development, and can be detected in single-cell mouse embryos (Sefton et al., 1992;

Vestweber et al., 1987). As determined in both mammalian and avian embryos, it is initially expressed in all three germ layers, but is generally lost from regions of ectoderm giving rise to neural derivatives and from mesoderm as it assumes a mesenchymal behavior (Thiery et al., 1984; Damjanov et al., 1986). It remains prominently expressed in ectodermal derivatives such as skin and in endodermal epithelia. Antibodies to E-cadherin have been shown to disrupt normal epidermal morphogenesis through blocking feather development (Gallin et al., 1986) and keratinocyte stratification (Wheelock and Jensen, 1992). Langerhans cells, the leukocytes of epidermis, have been suggested to localize and interact with epidermal cells through the expression of E-cadherin, since they express E-cadherin and anti-E-cadherin antibodies inhibit their interaction with keratinocytes (Tang et al., 1993). This is the first description of cadherin expression among leukocytes and indicates that cadherins may play important roles in the biology of a wider variety of cell types than previously demonstrated.

While most studies of cadherin expression during early development have focused on the embryo proper, the cadherins play a role in the development of extraembryonic structures as well. E-cadherin is expressed by human trophoblast cells during fusion and is down-regulated following their formation of syncitial trophoblast (Coutifaris et al., 1991). Syncitium formation was inhibited by anti-E-cadherin antibodies, suggesting a role for this protein in aggregation and fusion of trophoblast cells. P-cadherin is detected in extraembryonic tissues of implantation stage mouse embryos and subsequently in maternal uterine tissue, suggesting it may play a role in implantation (Nose and Takeichi, 1986).

C. N-cadherin Is Expressed in the Development of Numerous Tissues

As will be discussed, N-cadherin plays a dominant role in the development of the nervous system. However, N-cadherin has been found to be widely expressed during early stages of avian embryogenesis, in derivatives of all three germ layers (Duband et al., 1987, 1988; Hatta and Takeichi, 1986; Hatta et al., 1987). Expression was first detected on mesodermal cells which had migrated through the primitive streak, and many mesodermal derivatives including notochord, somites, nephrotome, and lateral plate all expressed N-cadherin, with polarized expression during development of epithelial organization, and diffuse distribution on mesenchymal cells. Changes in the expression of N-cadherin were found during the subsequent development of these structures, with the most striking changes occurring in concert with changes in cell segregation patterns, such as tubulation to form the Mullerian ducts, or epithelialization of somites. Among ectodermal derivatives the most prominent site of N-cadherin expression was the developing nervous system, with additional expression found in the developing otic, olfactory, and lens placodes. The switch from E-cadherin to N-cadherin expression in these ectodermal derivatives correlates with the segregation and involution of the neural plate and lens and otic placodes. Among endodermal derivatives, the visceral pouches expressed high levels of N-cadherin although most other endodermal derivatives expressed weakly if at all.

D. Cadherin Expression Is Similar But Not Identical Across Species

Comparison of studies of cadherin expression indicates that the observed patterns are often conserved but not identical across species. For example, in *Xenopus* development, E-cadherin expression is first detected in non-neural ectoderm of late gastrula stage embryos concommittantly with expression of N-cadherin in the neural plate (Levi et al., 1991a). At earlier stages, another cadherin, EP-cadherin, is found in all cells from oocytes to the late blastula stage (Levi et al., 1991b). However, EP-cadherin is lost from most of the cells contributing to the nervous system, while it continues to be expressed in non-neural ectoderm and endoderm where E-cadherin is now found as well. It is also expressed in the somites and later in muscle. P-cadherin, which is expressed at high levels in the mouse placenta, is not detected in the bovine placenta (Liaw et al., 1990). Immunoblotting studies with an antiserum recognizing N-cadherin in several species indicated that in general expression patterns are conserved between avian and murine tissues during development (Lagunowich et al., 1990). The exceptions to this cited here suggest that in some cases relative differences in cadherin expression are perhaps more important than the absolute expression of a particular cadherin for the segregation of tissues, and that some cadherins may substitute for one another to accomplish similar morphogenetic ends.

E. Perturbation Studies Shed Light on Additional Cadherin Functions

The earliest expression of N-cadherin in several embryonic systems occurs following neural induction when it is found expressed in the neural plate (Hatta and Takeichi, 1986). The role of N-cadherin in early development has been explored via both negative and positive changes in expression. By injection of N-cadherin mRNA into xenopus embryos, premature ectopic expression was obtained which led to the development of abnormal cell boundaries due to altered cell mixing and severe morphological defects, including abnormal development of the ectoderm and neural tube (Detrick et al., 1990; Fujimori et al., 1990). Negative perturbation of N-cadherin in early development has been carried out by the expression in vivo, in early *Xenopus* embryos, of a truncated form of N-cadherin lacking the external domains (Kintner, 1992). The construct was expressed by injection into blastomeres of mRNA encoding the transmembrane and cytoplasmic domains. The expressed protein was found to inhibit cell adhesion and to inhibit catenin binding to E-cadherin. In injected embryos, the integrity of the ectoderm was disturbed, and also led to abnormal gastrulation. Interestingly, a series of deletion constructs demonstrated that at least two distinct sites on the truncated protein were able to interfere with cadherin function. One of these corresponds to the site involved in catenin binding, while the other is closer to the membrane and may be involved in cadherin clustering.

F. Multiple Cadherins Are Expressed During Development of the Nervous System

Given the complexity of the nervous system, it is perhaps not surprising that it is the site of multiple cadherin expression. N-cadherin is prominently expressed in the neural tube from its formation until late stages in neural histogenesis, when its expression becomes more restricted. A study of N-cadherin expression during chick spinal cord development indicated that as in other CNS regions, N-cadherin is down-regulated, but remains expressed at high levels in the floor plate, in association with commissural segments of axons (Shiga and Oppenheim, 1991). At later stages of chick retinal development, N-cadherin expression is down-regulated and remains expressed at a high level only at the retinal outer limiting membrane, which is composed of adherens junctions (Matsunaga et al., 1988b; Lagunowich and Grunwald, 1989). While initially widely expressed in the chick embryo brain, during later phases of histogenesis expression becomes restricted to cells lining the ventricles and also in choroid plexus, most likely ependymal cells (Lagunowich et al., 1992). This study also demonstrated that N-cadherin expression remained high in the brainstem and spinal cord floorplate, a region which may serve as a guidepost for axonal migration during development.

The role of N-cadherin in chick retinal histogenesis was explored by incubation of retinal explants in vitro with antibodies directed against N-cadherin (Matsunaga et al., 1988b). While control explants developed the characteristic multilaminar appearance of retinal tissues, the antibody-treated tissues became disordered with abnormal laminations. This abnormal development was especially pronounced in the region of photoreceptors, perhaps reflecting the normally high expression of N-cadherin at the outer limiting membrane. However, there did not appear to be any effect on differentiation per se as determined by expression of photoreceptor marker antigens.

While N-cadherin is the predominant cadherin expressed in developing neural tissues, other cadherins have been described within the nervous system as well. Even E-cadherin, which had long been considered to not be expressed there, has been shown to be expressed transiently in restricted regions of the metencephalon, mesencephalon, diencephalon, and cerebellum, persisting longest in the roof plate (Shimamura and Takeichi, 1992). Explants of E-cadherin-positive brain regions, when treated with anti-E-cadherin antibodies, underwent alterations in morphology. E-cadherin is also expressed in sensory neurons of mouse dorsal root ganglia, satellite, and Schwann cells during development and into adulthood (Shimamura et al., 1992). Two interesting observations made as part of this study were that only a subset within each class of cell expressed E-cadherin, and that the target region of the spinal cord to which the E-cadherin-positive DRG neurons projected was restricted to a region of lamina II of the dorsal horn which extended all the way rostrally to include part of the trigeminal nucleus. This may indicate a specific role for this cadherin in establishment of sensory pathways in the developing CNS.

This interpretation is also supported by comparative studies of R- and N-cadherin expression. R-cadherin has recently been identified in chick embryos (Inuzuka et al., 1991a, 1991b). While retina and brain were identified as expressing the highest levels, it was found to be expressed, as is N-cadherin, in all neural tissues. Interestingly, the temporal pattern of expression of R- and N-cadherin are reciprocal, in that N-cadherin is expressed very early and is then down-regulated, while R-cadherin expression increase during development. In addition, expression of N- and R-cadherin even within tissues in which they are contemporaneously expressed was not identical. As with N-cadherin, R-cadherin is also not restricted to neural tissues, as it was found to be expressed in notochord, myotome, and early skeletal muscle. In a later study, the detailed expression of N- and R-cadherin were compared during the development of sensory and motor axon systems of the embryonic chick (Redies et al., 1992). It was observed that these molecules are often expressed in a reciprocal fashion, as in the hindbrain where N-cadherin is preferentially expressed on sensory axons while R-cadherin is found on motor axons. N- and R-cadherin may differ in other functional ways as well. The developmental loss of the in vitro histogenetic potential of dissociated retinal cells parallels the loss of N-cadherin expression (Grunwald et al., 1981; Matsunaga et al., 1988b; Lagunowich and Grunwald, 1989), even while R-cadherin expression increases. This suggests that N- and R-cadherin play very different roles in retinal histogenesis, or that other factors than cadherin expression become limiting as the retina differentiates.

N-cadherin expression is dynamically regulated during neural crest development, as revealed by the earlier cited antibody localization studies. More recently, the role of N-cadherin in neural crest development has been examined in the embryonic chick (Akitaya and Bronner-Fraser, 1992; Bronner-Fraser et al., 1992). In agreement with earlier studies, N-cadherin was detected on the neural folds prior to tube closure and on the closed neural tube. However, no immunoreactivity was detected on migrating neural crest cells which had left the neural tube, although it was expressed on neural crest cells during condensation into ganglia. The functional role of N-cadherin in this process was explored by microinjection of anti-N-cadherin antibodies into the cranial cavity. This led to the development of abnormal and misshapen or open neural tubes and the formation of ectopic aggregates of neural crest cells. The expression of T-cadherin suggests it may play a role in determining neural crest migration patterns (Ranscht and Bronner-Fraser, 1991).

Finally, in the nervous system, cadherins are expressed not only in nervous tissue proper, but also in associated structures. E-cadherin is found in arachnoid villi (Yamashima et al., 1992) and N- and B-cadherin are expressed in choroid plexus (Napolitano et al., 1991; Lagunowich et al., 1992).

G. Multiple Cadherins Are Expressed During Muscle Development

As mentioned previously, immunohistochemical studies of embryonic development indicated that N-cadherin is prominently expressed in the mesoderm of embryos and early in muscle development, but is down-regulated during later stages of myogenesis (Hatta et al., 1987). N-cadherin is expressed on the surface of avian skeletal myoblasts (Soler and Knudsen, 1991), and antibodies against N-cadherin inhibit myoblast fusion (Knudsen et al., 1990; Pouliot et al., 1990). Myoblast fusion is inhibited by synthetic peptides containing the cadherin HAV recognition sequence and with antibodies directed against N-cadherin (Mege et al., 1992). The more recently identified M-cadherin, also expressed in muscle, is found at low levels in myoblasts but is up-regulated during myotube formation (Donalies et al., 1991). Thus, it is likely that at least two cadherins contribute to the process of myogenesis, although two more, B- and T-cadherin, have been identified in muscle as well (Ranscht and Dours-Zimmerman, 1991; Napolitano et al., 1991). Interestingly, N-cadherin mRNA and protein expression in skeletal muscle appears to be under neural regulation, with down-regulation occurring upon innervation and re-expression occurring following denervation (Hahn and Covault, 1992). This study further indicated that the level of N-cadherin expressed in mature muscle fibers may depend in part on the muscle fiber type and on the level of motor activity.

In addition to its already discussed role in adherens junctions of heart cells, N-cadherin may play important roles in other aspects of cardiac development. Recent studies of N-cadherin expression indicate it may play a role in formation of the pericardial coelom, and in later stages is down-regulated in the endocardium but continues to be expressed in formation of the myocardium (Linask, 1992). A confocal scanning laser study of heart development indicated that N-cadherin may serve to align myofibrils between adjacent myocytes and play a role in the important morphogenetic event of cardiac looping (Shiraishi et al., 1993).

VIII. CADHERINS ARE INVOLVED IN FUNCTIONS BEYOND ADHESION SUCH AS INTRACELLULAR SIGNALING: STUDIES OF N-CADHERIN'S ROLE IN NEURITE GROWTH INDICATE A ROLE IN TRANSMEMBRANE SIGNALING

The prominent expression of N-cadherin during nervous system development has led to many studies of its expression and function during neurogenesis, with particular attention paid to its involvement in neurite growth. Many adhesion molecules are expressed on neurites, and the experimental abolition of neurite growth over complex substrates, such as other neurons or glial cells, has been shown to require simultaneous inhibition of several different classes of adhesion molecules. For example, antibody inhibition studies of ciliary ganglion cell neurite growth on astrocytes indicated that both cadherins and integrins were involved in process outgrowth (Tomaselli et al., 1988). Similar studies of growth of motor neuron axons over Schwann cells suggested that integrins, immunoglobulin gene superfamily members, as well as cadherins all functioned together in an additive fashion to promote neurite outgrowth (Bixby et al., 1988). Developmental changes in the relative roles of different adhesive receptors were also found in these studies, indicating that the promotion of neurite outgrowth is a highly context-dependent process exhibiting both cell-type and age-dependent specificity.

Studies using transfected fibroblastic cell lines showed that when such cells expressed N-cadherin they could serve as a substrate for neurite growth (Matsunaga et al., 1988a). Such transfected cells have been used to analyze aspects of the mechanism of cadherin-mediated neurite growth. Using a variety of neurons including retinal and cerebellar cells, the response to N-cadherin expressed by transfected 3T3 fibroblasts appeared to increase linearly with increased expression of N-cadherin, as opposed to the highly cooperative response to N-CAM, and the two adhesion molecules were found to potentiate each other, although the developmental profiles of response to these adhesion molecules differed (Doherty et al., 1991b). N-cadherin-expressing transfected fibroblasts were found to induce the differentiation and neurite extension of PC12 cells in a transcription-independent manner (Doherty et al., 1991a). This response was inhibited by pertussis toxin and calcium channel blockers, could be modulated by drugs which affect protein kinases, and occurred in a transcription-independent fashion. This appeared to be mediated by a pathway independent of that induced by nerve growth factor. These observations clearly indicate that transmembrane signalling events triggered by cell adhesion molecules can regulate aspects of neuronal differentiation. However, the extent to which this occurred independently of adhesive function per se remained in question. This was further examined using a similar experimental paradigm in which it was found that PC12 cells growing on a 3T3 monolayer could be induced to differentiate in a manner similar to that induced by N-cadherin by potassium-induced depolarization or agonist-induced activation of calcium channels (Saffel et al., 1992). Like the N-cadherin stimulated growth, the potassium-effect was blocked by calcium channel antagonists and kinase inhibitors, however, pertussis toxin had no effect. These results support the idea that N-cadherin may affect PC12 cells by activating second messenger systems rather than acting as an adhesion protein per se, and that the chain of events proceeds from adhesion receptor activation, via a G-protein-mediated process, to activation of calcium channels, and though a kinase-dependent step to affect neurite growth.

N-cadherin promotion of neurite growth may be modulated by a number of factors, such as incorporation of exogenous ganglioside GM1 into PC12 cells (Doherty et al., 1992a). Fibroblasts expressing N-cadherin and N-CAM can both serve as a substrate for neurite growth among hippocampal neurons, although over the course of development from late embryonic to early postnatal days responsiveness to N-CAM decreases while that for N-cadherin increases (Doherty et al., 1992b). Neurite outgrowth is also inhibited by antibodies against the N-acetylgalactosaminylphosphotransferase which, as previously, is functionally and structurally closely associated with N-cadherin (Gaya-Gonzalez et al., 1991).

While many of the above studies examine the function of N-cadherin in complex settings, such as when expressed by transfected cells, other studies have evaluated the biological activity of isolated N-cadherin. Purified N-cadherin was found to be a potent substrate for neurite growth when directly coated onto a nitrocellulose support and used as a substrate for ciliary ganglion neurons (Bixby and Zhang, 1990). This approach has permitted the further analysis of N-cadherin function, where it was found that as opposed to integrin-mediated neurite growth, which is inhibited by protein kinase inhibitors, N-cadherin-mediated outgrowth was stimulated by these drugs (Bixby and Jhabvala, 1990). However, this effect was transient, as the potentiation seemed to affect initiation of neurite growth rather than its maintenance, and the protein kinase agonist TPA was found to stimulate N-cadherin-mediated neurite growth. These results suggest that the involvement of intracellular signaling systems with adhesion molecule function may depend on the adhesion molecule as well as the stage in the process. Purified NCAD90, the naturally occuring soluble extracellular fragment of N-cadherin discussed earlier, also functions in a similar manner when immobilized on nitrocellulose, as a promoter of neuronal adhesion and neurite outgrowth (Paradies and Grunwald, 1993). Studies utilizing purified adhesion molecules including N-cadherin as adhesive substrates have examined growth cone behavior, with the conclusions that different adhesion molecules have different effects on growth cone morphology (Payne et al., 1992), and that relative adhesiveness per se does not appear to be the major determinant with respect to axonal pathway choice, growth rate or extent of fasciculation (Lemmon et al., 1992).

IX. ALTERATIONS OF CADHERIN EXPRESSION AND FUNCTION MAY CONTRIBUTE TO SEVERAL DISEASE PROCESSES

A. Cancer and Metastasis

Most studies of cadherins have been done in the context of embryonic development. However, cadherin expression persists in adult organisms and presumably plays important roles in the function and stability of mature tissues. It is therefore likely that misexpression of cadherins may be involved in a variety of pathologies. This has been examined most extensively in cancer cells, since the role of cadherins as adhesion molecules prompted examinations of their possible role in metastasis. This work has focused on E-cadherin and carcinomas. Several studies have indicated that an inverse correlation exists between expression of E-cadherin and cell invasiveness and metastasis. Among a large panel of human squamous cell skin carcinomas, the highly differentiated tumors expressed as much E-cadherin as normal epithelia, while E-cadherin levels were much reduced in the less well differentiated tumors (Schipper et al., 1991). Among tumor cells which had infiltrated lymph nodes, low E-cadherin expression was found. Similarly, decreased expression of E-cadherin has been associated with progression of rat prostatic cancer (Bussemakers et al., 1992), mouse skin cancer (Ruggeri et al., 1992), and human meningiomas (Tohma et al., 1992). Genetic elements which regulate E-cadherin expression are less active in poorly differentiated tumor cells than in well-differentiated cells (Behrens et al., 1991). While most attention has been focused on E-cadherin and cancer, one study of N-cadherin expression in tumor cells has been done, where expression of this neural cadherin was examined in a series of human retinoblastoma cell lines (Schiffman and Grunwald, 1992). This study indicated that N-cadherin is expressed by retinoblastoma cells, although both qualitative and quantitative variations in expression were observed between the various cell lines. Interestingly, while not absolute, a correlation existed between the level of expression and the extent of calcium-dependent adhesiveness and morphology of adhesions between the cells. However, it is important to note that the expression level of cadherins is not likely to be the only factor determining changes in the behavior of these cells. For example, the fibroblast growth factor-induced dispersion of rat bladder carcinoma cells, which results in their transition from an epithelial to mesenchymal behavior, results in a redistribution of cell surface E-cadherin without a reduction in its total expression or the calcium-dependent adhesiveness of the cells (Boyer et al., 1992). In addition, a line of human lung cancer cells with defective cell-cell adhesion was shown to express E-cadherin but to be defective in expression of α-catenin (Shimoyama et al., 1992). Finally, transformation of cells may result not only in the loss of expression of some cadherins, but the expression of innappropriate cadherins of a less mature phenotype, as in pancreatic tumors (Moller et al., 1992).

B. Other Pathological Conditions

Cadherin misexpression has been suggested to play a role in other pathogenetic processes. A possible role for cadherins has been suggested as the basis for epithelial abnormalities in hereditary hair loss in children (Baden et al., 1992). Cadherin misexpression has also been associated with Darier's and Hailey-Hailey diseases, which result in abnormal keratinocyte adhesion (Burge and Schomberg, 1992). A desmosomal cadherin called PVA has been identified as the target antigen in the autoimmune skin blistering disease pemphigus vulgaris (Amagai et al., 1991). Passive transfer of anti-PVA antibodies caused pemphigus-like disturbances in the skin of experimental mice (Amagai et al., 1992). As calcium-binding proteins, the cadherins are potential targets for calcium antagonists. Indeed it has been demonstrated that cadmium alters the distribution of E-cadherin and formation of cell junctions among renal epithelial cells (Prozialek and Niewenhuis, 1991).

X. CONCLUSION AND FUTURE DIRECTIONS

The cadherins have been identified as key molecules in the regulation of cell interactions during development and are now implicated in a number of pathological processes. While much has been learned in the last decade about cadherin structure and function, the growing diversity of the cadherin gene superfamily and the emerging complexity of regulatory mechanisms governing cadherin expression and function demonstrate that much remains to be discovered regarding the cadherins. How many cadherins are there? How is cadherin expression regulated genetically and epigenetically? Why do tissues express multiple cadherins and how do their functions differ? What types of information are communicated between cadherins and the cytoskeleton and via cadherin-mediated second messegner pathways? Can the analysis of cadherins be of diagnostic and prognostic use in clinical situations? These and other questions are and will continue to be addressed as more is learned about the biology of these fascinating molecules.

ACKNOWLEDGMENTS

The author's work is supported by grants EY06658 from the National Institutes of Health and BNS9021703 from the National Science Foundation.

REFERENCES

Akitaya, T., & Bronner-Fraser, M. (1992). Expression of cell adhesion molecules during initiation and cessation of neural crest migration. Dev. Dynamics 194, 12-20.

Amagai, M., Klaus-Kovtum, V., & Stanley, J.R. (1991). Autoantibodies against a novel epithelial cadherin in pemphigus vulgaris, a disease oc cell adhesion. Cell 67, 869-877.

Amagai, M., Karpati, S., Prussick, R., Klaus-Kovtun, V., & Stanley, J.R. (1992). Autoantibodies against the amino-terminal cadherin-like binding domain of pemphigus vulgaris antigen are pathogenic. J. Clin. Invest. 90, 919-926.

Baden, H.P., Kvedar, J.C., & Magro, C.M. (1992). Loose anagen hair as a cause of hereditary hair loss in children. Arch. Dermatol. 128, 1349-1353.

Balsamo, J., & Lilien, J. (1990). N-cadherin is stably associated with and is an acceptor for a cell surface N-acetylgalactosaminylphosphotransferase. J. Biol. Chem. 265, 2923-2928.

Balsamo, J., Pratt, R.S., Emmerling, M.R., Grunwald, G.B., & Lilen, J. (1986). Identification of the chick neural retina cell surface N-acetylgalactosaminyltransferase using monoclonal antibodies. J. Cell. Biochem. 32, 125-141.

Balsamo, J., Thiboldeaux, R., Swaminathan, N., Lilien, J. (1991). Antibodies to the retina N-acetylgalactosaminyl-phosphotransferase modulate N-cadherin -mediated adhesion and uncouple the N-cadherin transferase complex from the actin-containing cytoskeleton. J. Cell Biol. 113, 429-436.

Bauer, G.E., Balsamo, J., & Lilien, J. (1992). Cadherin-mediated adhesion in pancreatic islet cells is modulated by a cell surface N-acetylgalactosaminylphosphotransferase. J. Cell Sci. 103, 1235-1241.

Behrens, J., Birchmeier, W., Goodman, S.L., & Imhof, B.A. (1985). Dissociation of Madin-Darby canine kidney epithelial cells by the monoclonal antibody Arc-1: mechanistic aspects and identification of the antigen as a component related to uvomorulin. J. Cell Biol. 101, 1307-1315.

Behrens, J., Lowrick, O., Klein-Hitpass, L., & Birchmeier, W. (1991). The E-cadherin promoter: functional analysis of a GC-rich region and an epithelial cell-specific palindromic regulatory element. Proc. Natl. Acad. Sci. USA 88, 11495-11499.

Behrens, J., Vakaet, L., Friis, R., Winterhager, E., Van Roy, F., Mareel, M.M., & Birchmeier, W. (1993). Loss of epithelial differentiation and gain of invasiveness correlates with tyrosine phosphorylation of the E-cadherin/β-catenin complex in cells transformed with a temperature-sensitive v-SRC gene. J. Cell Biol. 120, 757-766.

Bertellotti, R., Rutishauser, U., & Edelman, G.M. (1980). A cell surface molecule involved in aggregation of embryonic liver cells. Proc. Natl. Acad. Sci. USA 77, 4831-4835.

Bixby, J.L. & Jhabvala, P. (1990). Extracellular matrix molecules and cell adhesion molecules induce neurites through different mechanisms. J. Cell Biol. 111, 2725-2732.

Bixby, J.L., & Zhang, R. (1990). Purified N-cadherin is a potent substrate for the rapid induction of neurite outgrowth. J. Cell Biol. 110, 1253-1260.

Bixby, J.L., Lilien, J., & Reichardt, L.F. (1988). Identification of the major proteins that promote neuronal process outgrowth on Schwann cell *in vitro*. J. Cell Biol. 107, 353-361.

Blaschuk, O.W., Sullivan, R., Davis, S., & Pouliot, Y. (1990). Identification of a cadherin cell adhesion recognition sequence. Dev. Biol. 139, 227-229.

Boller, K., Vestweber, D., & Kemler, R. (1985). Cell adhesion molecule uvomorulin is localized in the intermediate junctions of adult intestinal epithelial cells. J. Cell Biol. 100, 327-332.

Boyer, B., Dufour, S., & Thiery, J.P. (1992). E-cadherin expression during the acidic FGF-induced dispersion of a rat bladder carcinoma cell line. Exp. Cell Res. 201, 347-357.

Brackenbury, R., Rutihauser, U., & Edelman, G.M. (1981). Distinct calcium-independent and calcium-dependent adhesion systems of chicken embryo cells. Proc. Natl. Acad. Sci. USA 78, 387-391.

Bronner-Fraser, M., Wolf, J.J., & Murray, B.A. (1992). Effects of antibodies against N-cadherin and N-CAM on the cranial neural crest and neural tube. Dev. Biol. 153, 291-301.

Burge, S.M., & Schomberg, K.H. (1992). Adhesion molecules and related proteins in Darier's disease and Hailey-Hailey disease. British J. Dermatol. 127, 335-343.

Bussemakers, M.J.G., van Moorselaar, R.J.A., Giroldi, L.A., Ichikawa, T., Isaacs, J.T., Takeichi, M., Debruyne, F.M.J., & Schalken, J.A. (1992). Cancer Res. 52, 2916-2922.

Buxton, R.S., & Magee, A.I. (1992). Structure and interactions of desmosomal and other cadherins. Seminars in Cell Biol. 3, 157-167.

Byers, S., Amaya, E., Munro, S., & Blaschuk, O. (1992). Fibroblast growth factor receptors contain a conserved HAV region common to cadherins and influenza strain A hemmaglutinins: A role in protein-protein interactions? Dev. Biol. 152, 411-414.

Chen, B., Blaschuk, O.W., & Hales, B.F. (1991). Cadherin mRNAs during rat embryo development *in vivo* and *in vitro*. Teratology 44, 581-590.

Citi, S. (1992). Protein kinase inhibitors prevent junction dissociation induced by low extracellular calcium in MDCK epithelial cells. J. Cell Biol. 117, 169-178.

Cook, J.H., & Lilien, J. (1982). The accessibility of certain proteins on embryonic chick neural retina cells to iodination and tryptic removal is altered by calcium. J. Cell Sci. 55, 85-103.

Cook, J.H., Pratt, R.S., & Lilien, J. (1984). Turnover and orientation of the major neural retina cell surface protein protected from tryptic cleavage by calcium. Biochem. 23, 899-904.

Coutifaris, C., Kao, L.C., Sehdev, H.M., Chin, U., Babalola, G.O., Blaschuk, O.W., & Strauss, J.F. (1991). E-cadherin expression during the differentation of human trophoblasts. Development 113, 767-777.

Covault, J., Liu, Q., & El-Deeb, S. (1991). Calcium-activated proteolysis of intracellular domains in the cell adhesion molecules NCAM and N-cadherin. Molec. Brain. Res. 11, 11-16.

Crittenden, S.L., Rutishauser, U., & Lilien, J. (1988). Identification of two structural types of calcium-dependent adhesion molecules in the chicken embryo. Proc. Natl. Acad. Sci USA 85, 3464-3468.

Cunningham, B.A., Leutzinger, Y., Gallin, W.J., Sorkin, B.C., & Edelman, G.M. (1984). Linear organization of the liver cell adhesion molecule L-CAM. Proc. Natl. Acad. Sci. USA 81, 5787-5791.

Cunningham, B.A. & Edelman, G.M. (1990). Structure, expression and cell surface modulation of cell adhesion molecules. In: *Morphoregulatory Molecules* (Edelman, G.M., Cunningham, B.A., & Thiery, J.P., Eds.). John Wiley and Sons, New York, pp. 9-40.

Cyr, D.G., Balschuk, O.W., & Roabire, B. (1992). Identification and developmental regulation of cadherin messenger ribonucleic acids in the rat testis. Endocrinology 131, 139-145.

Damjanov, I., Damjanov, A., & Damsky, C.H. (1986). Developmentally regulated expression of the cell-cell adhesion glycoprotein call-CAM 120/80 in peri-implantation mouse embryos and extraembryonic membranes. Dev. Biol. 116, 194-202.

Damsky, C.H., Richa, J., Solter, D., Knudsen, K.A., & Buck, C.A. (1983). Identification and purification of a cell surface glycoprotein mediating intercellular adhesion in embryonic and adult tissues. Cell 34, 455-466.

DeMarais, A.A., & Moon, R.T. (1992). The armadillo homolgos catenin and plakoglobin are differentially expressed during early development of *Xenopus* laevis. Dev. Biol. 153, 337-346.

Detrick, R.J., Dickey, D., & Kintner, C.R. (1990). The effects of N-cadherin misexpression on morphogenesis in *Xenopus* embryos. Neuron 4, 493-506.

Doherty, P., Ashton, S.V., Moore, S.E., & Walsh, F.S. (1991a). Morphoregulatory activities of NCAM and N-cadherin can be accounted for by G protein-dependent activation of L- and N-type neuronal calcium channels. Cell 67, 21-33.

Doherty, P., Rowett, L.H., Moore, S.E., Mann, D.A., & Walsh, F.S. (1991b). Neurite outgrowth in response to transfected N-CAM and N-cadherin reveals fundamental differences in neuronal responsiveness to CAMs. Neuron 6, 247-258.

Doherty, P., Ashton, S.V., Skaper, S.D., Leon, A., & Walsh, F.S. (1992a). Ganglioside modulation of neural cell adhesion molecule and N-cadherin-dependent neurite outgrowth. J. Cell Biol. 117, 1093-1099.

Doherty, P., Skaper, S.D., Moore, S.E., Leon, A., & Walsh, F.S. (1992b). A developmentally regulated switch in neuronal responsiveness to NCAM and N-cadherin in the rat hippocampus. Development 115, 885-892.

Donalies, M., Cramer, M., Ringwald, M., & Starzinski-Powitz, A. (1991). Expression of M-cadherin, a member of the cadherin multigene family, correlates with differentiation of skeletal muscle cells. Proc. Natl. Acd. Sci. USA 88, 8024-8028.

Duband, J.L., Dufour, S., Hatta, K., Takeichi, M., Edelman, G.M., & Thiery, J.P. (1987). Adhesion molecules during somitogenesis in the avian embryo. J. Cell Biol. 104, 1361-1374.

Duband, J.L., Volberg, T., Sabanay, I., Thiery, J.P., & Geiger, B. (1988). Spatial and temporal distribution of adherens-junction-associated adhesion molecule A-CAM during avian embryogenesis. Development 103, 325-344.

Edelman, G.M., Murray, B.A., Mege, R.M., Cunningham, B.A., & Gallin, W.J. (1987). Cellular expression of liver and neural cell adhesion molecules after transfection with their cDNAs results in specific cell-cell binding. Proc. Natl. Acad. Sci. USA 84, 8502-8506.

Ferreira, M.C., Grunwald, G.B., & Paradies, N.E. (1993). Characterization of the endogenous retinal protease which catalyzes N-cadherin turnover. Soc. Neurosci. Abstr. 19, 459.

Friedlander, D.R., Mege, R.M., Cunningham, B.A., & Edelman, G.M. (1989). Cell sorting-out is modulated by both the specificity and amount of different cell adhesion molecules (CAMs) expressed on cell surfaces. Proc. Natl. Acad. Sci. USA 86, 7043-7047.

Fujimori, T., & Takeichi, M. (1993). Disruption of epithelial cell-cell adhesion by exogenous expression of a mutated nonfunctional N—cadherin. Molec. Biol. Cell 4, 37-47.

Fujimori, T., Miyatani, S., & Takeichi, M. (1990). Ectopic expression of N-cadherin perturbs histogenesis in *Xenopus* embryos. Development 110, 97-104.

Fujimori, T., & Takeichi, M. (1993). Disruption of epithelial cell-cell adhesion by exogenous expression of a mutated nonfunctional N-cadherin. Molec. Biol. Cell 4, 37-47.

Gallin, W.J., Edelman, G.M., & Cunningham, B.A. (1983). Characterization of L-CAM, a major cell adhesion molecule from embryonic liver cells. Proc. Natl. Acad. Sci. USA 80, 1038-1042.

Gallin, W.J., Chuong, C.M., Finkel, L.H., & Edelman, G.M. (1986). Antibodies to liver cell adhesion molecule perturb inductive interactions and alter feather pattern and structure. Proc. Natl. Acad. Sci. USA 83, 8235-8239.

Gallin, W.J., Sorkin, B.C., Edelman, G.M., & Cunningham, B.A. (1987). Sequence analysis of a cDNA clone encoding the liver cell adhesion molecule, L-CAM. Proc. Natl. Acad. Sci. USA 84, 2808-2812.

Gaya-Gonzalez, L., Balsamo, J., Swaminatham, N., and Lilien, J. (1991). Antibodies to the retina N-acetylgalactosaminylphosphotransferase inhibit neurite outgrowth. J. Neurosci. Res. 29, 474-480.

Geiger, B., & Ayalon, O. (1992). Cadherins. Annu. Rev. Cell Biol. 8, 307-332.

Geiger, B., Volberg, T., Sabanay, I., & Volk, T. (1990). A-CAM: An adherens junction-specific cell adhesion molecule. In: *Morphoregulatory Molecules,* (Edelman, G.M., Cunningham, B.A., & Thiery, J.P. (Eds.). John Wiley and Sons, New York, pp. 57-79.

Geiger, B., Salomon, D., Takeichi, M., & Hynes, R.O. (1992). A chimeric N-cadherin/β1-integrin receptor which localizes to both cell-cell and cell-matrix adhesisions. J. Cell Sci. 103, 943-951.

Ginsburg, D., DeSimone, D., & Geiger, B. (1991). Expression of a novel cadherin (EP-cadherin) in unfertilized eggs and early Xenopus embryos. Development 111, 315-325.

Grunwald, G.B. (1991). The conceptual and experimental foundations of vertebrate embryonic cell adhesion research. In: *A Conceptual History of Modern Embryology,* (Gilbert, S.F., Ed.). Plenum, New York, pp. 129-158.

Grunwald, G.B. (1992). Cell adhesion and recognition in development. In: *Fundamentals of Medical Cell Biology, Vol. 7, Developmental Biology,* (Bittar, E.E., Ed.). JAI Press, Greenwich, CT, pp. 103-132.

Grunwald, G.B. (1993). Progress in the structural and functional analysis of cadherins. Curr. Opin. Cell Biol. 5, 797-805.

Grunwald, G.B., Geller, R.L., & Lilien, J. (1980). Enzymatic dissection of embryonic cell adhesive mechanisms. J. Cell Biol. 85, 766-776.

Grunwald, G.B., Bromberg, R.E.M., Crowley, N.J., & Lilien, J. (1981). Enzymatic dissection of embryonic cell adhesive mechanisms. II. Developmental regulation of an endogenous adhesive system in the chick neural retina. Dev. Biol. 86, 327-338.

Grunwald, G.B., Pratt, R.S., & Lilien, J. (1982). Enzymatic dissection of embryonic cell adhesive mechanisms. III. Immunological identification of a component of the calcium-dependent adhesive system of embryonic chick neural retina cells. J. Cell Sci. 55, 69-83.

Grunwald, G.B., Chagin, K., & Paradies, N.E. (1993). Some adhesive interactions mediated by N-cadherin can occur without the cytoplasmic domain and in the absence of calcium. Soc. Neurosci. Abstr. 19, 459.

Gumbiner, B., Stevenson, B., & Grimaldi, A. (1988). The role of the cell adhesion molecule uvomorulin in the formation and maintenance of the epithelial junctional complex. J. Cell Biol. 107, 1575-1587.

Hahn, C., & Covault, J. (1992). Neural regulation of N-cadherin gene expression in developing and adult skeletal muscle. J. Neurosci. 12, 4677-4687.

Hamaguchi, M., Matsuyoshi, N., Ohnishi, Y., Gotoh, B., Takeichi, M., & Nagai, Y. (1993). p60v-src causes tyrosine phosphorylation and inactivation of the N-cadherin-catenin cell adhesion system. EMBO J. 12, 307-314.

Hatta, K., & Takeichi, M. (1986). Expression of N-cadherin adhesion molecules associated with early morphogenetic events in chick development. Nature 320, 447-449.

Hatta, K., Okada, T.S., & Takeichi, M. (1985). A monoclonal antibody disrupting calcium-dependent cell-cell adhesion of brain tissues: Possible role of its target antigen in animal pattern formation. Proc. Natl. Acad. Sci. USA 82, 2789-2793.

Hatta, K., Takagi, S., Fujisawa, H., & Takeichi, M. (1987). Spatial and temporal expression pattern of N-cadherin cell adhesion molecules correlated with morphogenetic processes of chicken embryos. Dev. Biol. 120, 215-227.

Hatta, K., Nose, A., Nagafuchi, A., & Takeichi, M. (1988). Cloning and expression of cDNA encoding a neural calcium-dependent cell adhesion molecule: Its identity in the cadherin gene family. J. Cell Biol. 106, 873-881.

Hatta, M., Miyatani, S., Copeland, N.G., Gilbert, D.J., Jenkins, N.A., & Takeichi, M. (1991). Genomic organization and chromosomal mapping of the mouse P-cadherin gene. Nucleic Acids. Res. 19, 4437-4441.

Herrenknecht, K., Ozawa, M., Eckerskorn, C., Lottspeich, F., Lenter, M., & Kemler, R. (1991). The uvomorulin-anchorage protein α-catenin is a vinculin homologue. Proc. Natl. Acad. Sci. USA 88, 9156-9160.

Hirano, S., Nose, A., Hatta, K., Kawakami, A., & Takeichi, M. (1987). Calcium-dependent cell-cell adhesion molecules (cadherins): Subclass specificities and possible involvement of actin bundles. J. Cell Biol. 105, 2501-2510.

Hirano, S., Kimoto, N., Shimoyama, Y., Hirohashi, S., & Takeichi, M. (1992). Identification of a neural α-catenin as a key regulator of cadherin function and multicellular organization. Cell 70, 293-301.

Holtfreter, J. (1939). Tissue affinity, a means of embryonic morphogenesis. Arch. Exp. Zellforsch. 23, 169-209. (English translation in: *Foundations of Experimental Embryology* (1964), (Willier, B.H., & Oppenheimer, J.M., Eds.). Prentice-Hall, Englewood Cliffs, NJ, pp. 186-225).

Hyafil, F., Morello, D., Babinet, C., & Jacob, F. (1980). A cell surface glycoprotein involved in the compaction of embryonic carcinoma cells and cleavage stage embryos. Cell 21, 927-934.

Inuzuka, H., Miyatani, S., & Takeichi, M. (1991a). R—cadherin: a novel Ca ^{2}+ -dependent cell-cell adhesion molecule expressed in the retina. Neuron 7, 69-79.

Inuzuka, H., Redies, C., & Takeichi, M. (1991b). Differential expression of R- and N-cadherin in neural and mesodermal tissues during early chicken development. Development 113, 959-967.

Jensen, P.J., & Wheelock, M.J. (1992). Regulation of urokinase plasminogen activator localization in keratinocytes by calcium ion and E-cadherin. Exp. Cell Res. 202, 190-198.

Kaupmann, K., Becker-Follman, J., Schere, G., Jockosch, H., & Starzinski-Powitz, A. (1992). The gene for the cell adhesion molecule M-cadherin maps to mouse chromosome * and human chromosome 16q24.1-qter and is near the E-cadherin (uvomorulin) locus in both species. Genomics 14, 488-490.

Kemler, R. (1992). Classical cadherins. Sem. in Cell Biol. 3, 149-155.

Kemler, R., Babinet, C., Eisen, H., & Jacob, F. (1977). Surface antigen is early differentiation. Proc. Natl. Acad. Sci. USA 74, 4449-4452.

Kemler, R., Gossler, A., Mansouri, A., & Vestweber, D. (1990). The cell adhesion molecule uvomorulin. In: *Morphoregulatory Molecules,* (Edelman, G.M., Cunningham, B.A., & Thiery, J.P., Eds.). John Wiley and Sons, New York, pp. 41-56.

Kintner, C. (1992). Regulation of embryonic cell adhesion by the cadherin cytoplasmic domain. Cell 69, 225-236.

Knudsen, K.A., & Wheelock, M.J. (1992). Plakolgobin, or an 83-kD homologue distinct from β-catenin, interacts with E-cadherin and N-cadherin, J. Cell Biol. 118, 671-679.

Knudsen, K.A., McElwee, S.A., & Meyers, L. (1990). A role for Ca2+ dependent cell adhesion molecule, N—cadherin, in myoblast interaction during myogenesis. Dev. Biol. 138, 159-168.

Lagunowich, L.A., & Grunwald, G.B. (1989). Expression of calcium-dependent cell adhesion during ocular development: a biochmeical, histochemical and functional analysis. Dev. Biol. 135, 158-171.

Lagunowich, L.A., & Grunwald, G.B. (1991). Tissue and age-specificity of post-translational modification sof N-cadherin during chick embryo development. Differentiation 47, 19-27.

Lagunowich, L.A., Donoso, L.A., & Grunwald, G.B. (1990). Identification of mammalian and invertebrate analogues of the avian calcium-dependent cell adhesion protein N-cadherin with synthetic peptide directed antibodies against a conserved cytoplasmic domain. Biochem. Biophys. Res. Commun. 172, 313-320.

Lagunowich, L.A., Schneider, J.C., Chasen, S., & Grunwald, G.B. (1992). Immunohistochemical and biochemical analysis of N-cadherin expression during CNS development. J. Neurosci. Res. 32, 202-208.

Lee, M.M., & Grunwald, G.B. (1993). Biochemical analysis of N-cadherin phosphorylation in developing chick embryo tissues. Soc. Neurosci. Abstr. 19, 460.

Legan, P.K., Collins, J.E., & Garrod, D.R. (1992). The molecular biology of desmosomes and hemidesmosomes: What is in a name? Bioessays 14, 385-393.

Lemmon, V., Burden, S.M., Payne, H.R., Elmslie, G.J., & Hlavin, M.L. (1992). Neurite growth on different substrates: Permissive versus instructive influences and the role of adhesive strength. J. Neurosci. 12, 818-826.

Levi, G., Gumbiner, B.M., & Thiery, J.P. (1991a). The distribution of E-cadherin during Xenopus laevis development. Development 111, 145-158.

Levi, G., Ginsburg, D., Girault, J.M., Sabanay, I., Thiery, J.P., & Geiger, B. (1991b). EP-cadherin in muscles and epithelia of Xenopus laevis embryos. Development 113, 1335-1344.

Liaw, C.W., Cannon, C., Power, M.D., Kiboneka, P.K., & Rubin, L.L. (1990). Identification and cloning of two species of cadherins in bovine endothelial cells. EMBO J. 9, 2701-2708.

Linask, K. (1992). N-cadherin localization in early heart development and polar expression of $Na+$,$K+-$ATPase, and integrin during pericardial coelom formation and epithelialization of the differentiating myocardium. Dev. Biol. 151, 213-224.

Magnani, J.L., Thomas, W.A., & Steinberg, M.S. (1981). Two distinct adhesion mechanisms in embryonic chick neural retina cells. I. A kinetic analysis. Dev. Biol. 81, 96-105.

Mahoney, P.A., Weber, U., Onofrechuk, P., Biessman, H., Bryant, P.J., & Goodman, C.S. (1991). The fat tumor suppressor gene in *Drosophila* encodes a novel member of the cadherin gene superfamily. Cell 67, 853-868.

Matsunaga, M., Hatta, K., Nagafuchi, A., & Takeichi, M. (1988a). Guidance of optic nerve fibers by N-cadherin adhesion molecules. Nature 334, 62-64.

Matsunaga, M., Hatta, K., & Takeichi, M. (1988b). Role of N-cadherin cell adhesion molecules in the histogenesis of neural retina. Neuron 1, 289-295.

Matsuyoshi, N., Hamaguchi, M., Taniguchi, S., Nagafuchi, A., Tsukita, S., & Takeichi, M. (1992). Cadherin-mediated cell-cell adhesion is perturbed by v-src tyrosine phosphorylation in metastatic fibroblasts. J. Cell Biol. 118, 703-714.

McCrea, P.D., Turck, C.W., & Gumbiner, B. (1991). A homolog of the armadillo protein in Drosophila (plakolgobin) associated with E-cadherin. Science 254, 1359-1361.

McNeill, H., Ozawa, M., Kemler, R., & Nelson, W.J. (1990). Novel function of the cell adhesion molecule uvomorulin as an inducer of cell surface polarity. Cell 62, 309-316.

Mege, R.M., Matsuzaki, R., Gallin, W.J., Goldberg, I., Cunningham, B.A., & Edelman, G.M. (1988). Construction of epithelioid sheets by transfection of mouse sarcoma cells with cDNAs for chicken cell adhesion molecules. Proc. Natl. Acad. Sci USA 85, 7274-7278.

Mege, R.M., Goudou, D., Diaz, C., Nicolet, M., Garcia, L., Geraud, G., & Rieger, F. (1992). N-cadherin and N-CAM in myoblast fusion: Compared localization and effect of bloackade by peptides and antibodies. J. Cell Sci. 103, 897-906.

Meyer, R.A., Laird, D.W., Revel, J.P., & Johnson, R.J. (1992). Inhibition of gap junction and adherems junction assemsbly by connexin and A-CAM antibodies. J. Cell Biol. 119, 179-189.

Miyatani, S., Shimamura, K., Hatta, M., Nagafuchi, A., Nose, A., Matsunaga, M., Hatta, K., & Takeichi, M. (1989). Neural cadherin: Role in selective cell-cell adhesion. Science 245, 631-635.

Miyatani, S., Copeland, N.G., Gilbert, D.J., Jenkins, N.A., & Takeichi, M. (1992). Genomic structure and chromosomal mapping of the mouse N-cadherin gene. Proc. Natl. Acad. Sci. USA 89, 8443-8447.

Moller, C.J., Christgau, S., Willaimson, M.R., Madsen, O.D., Zhan-Po, N., Bock, E., & Baekkesov, S. (1992). Differential expression of neural cells adhesion molecule and cadherins in pancreatic islets, glucagonomas, and insulinomas. Molec. Endocrinol. 6, 1332-1342.

Moscona, A. (1961). Rotation-mediated histogenetic aggregation of dissociated cells. Exp. Cell Res. 22, 455-475.

Moscona, A. (1962). Analysis of cell recombinations in experimental synthesis of tissues in vitro. J. Cell Comp. Physiol. 60 (Suppl. 1), 65-80.

Nagafuchi, A., & Takeichi, M. (1988). Cell binding function of E-cadherin is regulated by the cytoplasmic domainn. EMBO J. 7, 3679-3684, 1988.

Nagafuchi, A., & Takeichi, M. (1989). Transmembrane control of cadherin-mediated cell adhesion: A 94 kDa protein functionally associated with a specific region of the cytoplasmic domain of E-cadherin. Cell Reg. 1, 37-44.

Nagafuchi, A., Shirayoshi, Y., Okazaki, K., Yasuda, K., & Takeichi, M. (1987). Transformation of cell adhesion properties by exogenously introduced E-cadherin. Nature 329, 341-343.

Nagafuchi, A., Takeichi, M., & Tsukita, S. (1991). The 102 kd cadherin-associated protein: Similarity to vinculin and posttranscriptional regulation of expression. Cell 65, 849-857.

Napolitano, E.W., Venstrom, K., Wheeler, E.F., & Reichardt, L.F. (1991). Molecular cloning and characterization of B-cadherin, a novel chick cadherin. J. Cell Biol. 113, 893-905.

Nelson, W.J., Shore, E.M., Wang, A.Z., & Hammerton, R.W. (1990). Identification of a membrane-cytoskeletal complex containing the cell adhesion molecule uvomorulin (E-cadherin), ankyrin, and fodrin in Madin-Darby canine kidney epithelial cells. J. Cell Biol. 107, 2377-2387.

Nose, A., & Takeichi, M. (1986). A novel cadherin adhesion moleclue: Its expression patterns associated with implantation and organogenesis of mouse enbryos. J. Cell Biol. 103, 2649-2658.

Nose, A., Nagafuchi, A., & Takeichi, M. (1987). Isolation of placental cadherin cDNA: Identification of a novel gene family of cell-cell adhesion molecules. EMBO J. 6, 3655-3661.

Nose, A., Tsuji, K., & Takeichi, M. (1990). Localization of specificity determining sites in cadherin cell adhesion molecules. Cell 61, 147-155.

Ozawa, M., & Kemler, R. (1990). Correct proteolytic cleavage is required for the cell adhesive function of uvomorulin. J. Cell Biol. 111, 1645-1650.

Ozawa, M., & Kemler, R. (1992). Molecular organization of the uvomorulin-catenin complex. J. Cell Biol. 116, 989-996.

Ozawa, M., Baribault, H., & Kemler, R. (1989). The cytoplasmic domain of the cell adhesion molecular uvomorulin associates with three independent proteins structurally related in different species. EMBO J. 8, 1711-1717.

Ozawa, M., Engel, J., & Kemler, R. (1990a). Single amino acid substitutions in one Ca^2+ binding site of uvomorulin abolish the adhesive function. Cell 63, 1033-1038.

Ozawa, M., Hoschutzky, H., Herrenknecht, K., & Kemler, R. (1990b). A possible new adhesive site in the cell adhesion molecule uvomorulin. Mech. Dev. 33, 49-56.

Paradies, N., & Grunwald, G.B. (1993). Purification and characterization of NCAD90, a soluble endogenous form of N-cadherin, which is generated by proteolysis during retinal development and retains adhesive and neurite-promoting function. J. Neurosci. Res. 36, 33-45.

Parker, A.E., Wheeler, G.N., Arnemann, J., Pidsley, S.C., Ataliotis, P., Thomas, C.L., Rees, D.A., Magee, A.I., & Buxton, R.S. (1991). Desmosomal glycoproteins II and III: Cadherin-like junctional molecules generated by alternative splicing. J. Biol. Chem. 266, 10438-10445.

Payne, H.R., Burden, S.M., & Lemmon, V. (1992). Modulation of growth cone morphology by substrate-bound adhesion molecules. Cell Motil. & Cytoskel. 21, 65-73.

Peifer, M., McCrea, P.D., Green, K.J., Wieschaus, E., & Gumbiner, B.M. (1992). The vertebrate adhesive junction proteins β-catenin and plakoglobin and the *Drosophila* segment polarity gene armadillo form a multigene family with similar properties. J. Cell Biol. 118, 681-691.

Peyrieras, N., Hyafil, F., Louvard, D., Ploegh, H.L., & Jacob, F. (1983). Uvomorulin: A nonintegral membrane protein of early mouse embryos. Proc. Natl. Acad. Sci. USA 80, 6274-6277.

Pouliot, Y. (1992). Phylogenetic analysis of the cadherin superfamily. Bioessays 14, 743-748.

Pouliot, Y., Holland, P.C., & Blaschuk, O.W. (1990). Developmental regulation of a cadherin during the differentiation of skeletal myoblasts. Dev. Biol. 141, 292-298.

Prozialek, W.C., & Niewenhuis, R.J. (1991). Cadmium (Cd^2+) disrupts Ca^2+-dependent cell-cell junctions and alters the pattern of E-cadherin immunolfuoresence in $LLC\text{-}PK_1$ cells. Biochem. Biophys. Res. Commun. 181, 1118-1124.

Ranscht, B. (1991). Cadherin cell adhesion molecules in vertebrate neural development. Seminars in the Neurosciences 3, 285-296.

Ranscht, B., & Bronner-Fraser, M. (1991). T-cadherin expression alternates with migrating neural crest cells in the trunk of the avian embryo. Development 111, 15-22.

Ranscht, B., & Dours-Zimmerman, M.T. (1991). T-cadherin, a novel cadherin cell adhesion molecule in the nervous system lacks the conserved cytoplasmic region. Neuron 7, 391-402.

Redies, C., Inuzuka, H., & Takeichi, M. (1992). Restricted expression of N- and R-cadherin on neurites of the developing chicken CNS. J. Neurosci. 12, 3525-3534.

Reid, R.A., & Hemperley, J.J. (1991). Human N-cadherin: Nucleotide and deduced amino acid sequence. Nucleic Acids Res. 18, 5896.

Ringwald, M., Schuh, R., Vestweber, D., Eistetter, H., Lottspeich, F., Engel, J., Dolz, R., Jahnig, F., Epplen, J., Mayer, S., Muller, C., & Kemler, R., (1987). The structure of cell adhesion molecule uvomorulin. Insights into the molecular mechanism of Ca^2+ -dependent cell adhesion. EMBO J. 6, 3647-3653.

Ringwald, M., Baribault, H. Schmidt, C., & Kemler, R. (1991). The structure of the gene coding for the mouse cell adhesion molecule uvomorulin. Nucl. Acids. Res. 19, 6533-6539.

Roark, E.F., Paradies, N.E., Lagunowich, L.A., & Grunwald, G.B. (1992). Evidence for endogenous proteases, mRNA level and insulin as multiple mechanisms of N-cadherin down-regulation during retinal development. Development 114, 973-984.

Ruggeri, B., Caamano, J., Slaga, T.J., Conti, C.J., Nelson, W.J., & Klein-Szanto, A.J.P. (1992). Altcrations in the expression of uvomorulin and Na+ ,K+—adenosine triphosphatase during mouse skin tumor progression. Am. J. Path. 140, 1179-1185.

Rutishauser, U., Acheson, A., Hall, A.K., Mann, D.M., & Sunshine, J. (1988). The neural cell adhesion molecule (NCAM) as a regulator of cell-cell interactions. Science 240, 53-57.

Saffel, J.L., Walsh, F.S., & Doherty, P. (1992). Direct activation of second messenger pathways mimics cell adhesion molecule-dependent neurite outgrowth. J. Cell Biol. 118, 663-670.

Salomon, D., Ayalon, O., Patel-King, R., Hynes, R., & Geiger, B. (1992). Extrajunctional distribution of N-cadherin in cultured human endothelial cells. Development 102, 7-17.

Schiffman, J.S., & Grunwasld, G.B. (1992). Differential expression of calcium-dependent cell adhesion in human retinoblastoma cell lines. Invert. Ophthalmol. Vis. Sci. 33, 1568-1574.

Schipper, J.H., Frixen, U.H., Behrens, J., Unger, A., Jahnke, K., & Birchmeier, W. (1991). E-cadherin expression in squamous cell carcinomas of head and neck: Inverse correlation with tumor dedifferentiation and lymph node metastasis. Cancer Res. 51, 6328-6337.

Schneider, R. (1992). The human protooncogene ret: A communicative cadherin? Trends in Biochem. Sci. 17, 468-469.

Schuh, R., Vesweber, D., Riede, I., Ringwald, M., Rosenberg, U.B., Jackle, H., & Kemler, R. (1986). Molecular cloning of the mouse cell adhesion molecule uvomorulin: cDNA contains a B1-related sequence. Proc. Natl. Acad. Sci. USA 83, 1364-1368.

Sefton, M., Johnson, M.H., & Clayton, L. (1992). Synthesis and phosphorylation of uvomorulin during mouse early development. Development 115, 313-318.

Shiga, T., & Oppenheim, R.W. (1991). Immunolocalization studies of putative guidance molecules used by axons and growth cones of intersegmental internurons in the chick embryo spinal cord. J. Comp. Neurol. 310, 234-252.

Shimamura, K., & Takeichi, M. (1992). Local and transient expression of E-cadherin involved in mouse embryonic brain morphogenesis. Development 116, 1011-1019.

Shimamura, K., Takahashi, T., & Takeichi, M. (1992). E-cadherin expression in a particular subset of sensory neurons. Dev. Biol. 152, 242-254.

Shimoyama, Y., Nagafuchi, A., Fujita, S., Gotoh, M., Takeichi, M., Tsukita, S., & Hirohashi, S. (1992). Cadherin dysfunction in a human cancer cell line: Possible involvement of loss of α-catenin expression in reduced cell-cell adhesiveness. Cancer Res. 52, 5770-5774.

Shiraishi, I., Takamatsu, T., & Fujita, S. (1993). 3-D observation of N-cadherin expression during cardiac myofibrillogenesos of the chick embryo using a confocal laser scanning microscope. Anat. Embryol. 102, 115-120.

Shirayoshi, Y., Nose, A., Iwasaki, K., & Takeichi, M. (1986). N-linked oligosaccharides are not involved in the function of a cell-cell binding glycoprotein E-cadherin. Cell Struct. Func. 11, 245-252.

Shore, E.M., & Nelson, W.J. (1991). Biosynthesis of the cell adhesion molecule uvomorulin (E-cadherin) in Madin-Darby canine kidney epithelial cells. J. Biol. Chem. 266, 19672-19680.

Simamura, K., & Takeichi, M. (1992). Local and transient expression of E-cadherin involved in mouse embryonic brain morphogenesis. Development 116, 1011-1019.

Simonneau, L., Broders, F., & Thiery, J.P. (1992). N-cadherin transcripts in *Xenopus laevis* from early tailbud to tadpole. Dev. Dynamics 194, 247-260.

Soler, A.P., & Knudsen, K.A. (1991). Colocalization of N-CAM and N-cadherin in avian skeletal myoblasts. Dev. Biol. 148, 389-392.

Sorkin, B.C., Hemperley, J.J., Edelman, G.M., & Cunningham, B.A. (1988). Structure of the gene for the liver cell adhesion molecule, L-CAM. Proc. Natl. Acad. Sci. USA 85, 7617-7621.

Sorkin, B.C., Gallin, W.J., Edelman, G.M., & Cunningham, B.A. (1991). Genes for two calcium-dependent cell adhesion molecules have similar structures and are arranged in tandem in the chicken genome. Proc. Natl. Acad. Sci. USA 88, 11545-11549.

Steinberg, M.S. (1963). Reconstruction of tissues by dissociated cells. Science 141, 401-408.

Steinberg, M.S., Armstrong, P.B., & Granger, R.E. (1973). On the recovery of adhesiveness by trypsin-dissociated cells. J. Membrane Biol. 13, 97-128.

Suzuki, S., Sano, K., & Tanihara, H. (1991). Diversity of the cadherin family: Evidence for eight new cadherins in nervous tissue. Cell Reg. 2, 261-270.

Takeichi, M. (1977). Functional correlation between cell adhesive properties and some cell surface proteins. J. Cell Biol. 75, 464-474.

Takeichi, M. (1988). The cadherins: Cell-cell adhesion molecules controlling animal morphogenesis. Development 102, 639-655.

Takeichi, M. (1990). Cadherins: A molecular family important in selective cell-cell adhesion. Annu. Rev. Biochem. 59, 237-252.

Takeichi, M. (1991). Cadherin cell adhesion receptors as a morphogenetic regulator. Science 251, 1451-1455.

Tang, A., Amagi, M., Granger, L.G., Stanley, J.R., & Udey, M.C. (1993). Adhesion of epidermal Langerhans cells to keratinocytes mediated by E-cadherin. Nature 361, 82-85.

Thiery, J.P., Delouvee, A., Gallin, W.J., Cunningham, B.A., & Edelman, G.M. (1984). Ontogenic expression of cell adhesion molecules: L-cam is found in epithelia derived from the three primary germ layers. Dev. Biol. 102, 61-78.

Tohma, Y., Yamashima, T., & Yamashita, J. (1992). Immuohistochemical localization of cell adhesion molecule epithelial cadherin in human arachnoid villi and meningiomas. Cancer Res. 52, 1981-1987.

Tomaselli, K.J., Neugebauer, K.M., Bixby, J.L., Lilien, J., & Reichardt, L.F. (1988). N-cadherin and integrins: Two receptor systems that mediate neuronal process outgrowth on astrocyte surfaces. Neuron 1, 33-43.

Townes, P.L., & Holtfreter, J. (1955). Directed movements and selective adhesion of embryonic amphibian cells. J. Exp. Zool. 128, 53-120.

Urishihara, H., Pzaki, H.S., & Takeichi, M. (1979). Immunological detection of cell surface components related with aggregation of chinese hamster and chick embryonic cells. Dev. Biol. 70, 206-216.

Vestal, D., & Ranscht, B. (1992). Glycosyl phosphatidulinositol-anchored T-cadherin mediates calcium-dependent, homophilic adhesion. J. Cell Biol. 119, 451-461.

Vestweber, D., & Kemler, R. (1984). Some structural and functional spects of cell adhesion molecule uvomorulin. Cell Differ. 15, 269-273.

Vestweber, D., Gossler, A., Boller, K., & Kemler, R. (1987). Expression and distribution of cell adhesion molecule uvomorulin in mouse preimplantation embryos. Dev. Biol. 124, 451-456.

Volberg, T., Geiger, B., Dror, R., & Zick, Y. (1991). Modulation of intercellular adherens-type junctions and tyrosine phosphorylation of their components in RSV transformed cultured chick lens cells. Cell Regul. 2, 105-120.

Volk, T., & Geiger, B. (1984). A 135-kd membrane protein of intercellular adherens junctions. EMBO J. 3, 2249-2260.

Volk, T., & Geiger, B. (1986). A-CAM: A 135kd receptor of intercellular adherens junction. I. Immunoelectron microscopic localization and biochemical studies. J. Cell Biol. 103, 1441-1450.

Volk, T., Cohen, O., & Geiger, B. (1987). Formation of heterotypic adherens-type junctions between L-CAM-containing liver cells and A-CAM-containing lens cells. Cell 50, 987-994.

Volk, T., Volberg, T., Sabanay, I., & Geiger, B. (1990). Cleavage of A-CAM by endogenous proteinases in cultured lens cells and in developing chick embryos. Dev. Biol. 139, 314-326.

Wagner, A.P., Beck, K.D., & Reck, G. (1992). Neural cell adhesion molecule (NCAM) and N-cadherin mRNA during development and aging: Selective reduction in the 7.4 kb and 6.7 kb NCAM mRNA levels in the hippocampus of adult rats. Mech. Aging & Dev. 62, 201-208.

Walsh, F.S., Barton, C.H., Putt, W., Moore, S.E., Kelsell, D., Spurr, N., & Goodfellow, P.N. (1990). N-cadherin gene maps to human chromosome 18 and is not linked to the E-cadherin gene. J. Neurochem. 55, 805-812.

Wheelock, M.J. (1990). Catenin association with E-cadherin changes with the state of polarity of HT-29 cells. Exp. Cell Res. 191, 186-193.

Wheelock, M.J., & Jensen, P.J. (1992). Regulation of keratinocyte intercellular junction organization and epidermal morphogenesis by E-cadherin. J. Cell Biol. 117, 415-425.

Wheelock, M.J., & Knudsen, K.A. (1991). N-cadherin-associated proteins in chicken muscle. Differentiation 46, 35-42.

Wheelock, M.J., Buck, C.A., Bechtol, K.B., & Damsky, C.H. (1987). Soluble 80-kd fragment of cell-CAM 120/80 disrupts cell-cell adhesion. J. Cell. Biochem. 34, 187-202.

Winkel, G.K., Ferguson, J.E., Takeichi, M., & Nuccitelli, R. (1990). Activation of protein kinase C triggers premature compaction in the four-cell stage mouse embryo. Dev. Biol. 138, 1-15.

Wollner, D.A., Krzeminski, K.A., & Nelson, W.J. (1992). Remodelling the cell surface distribution of membrane proteins during the development of epithelial cell polarity. J. Cell Biol. 116, 889-899.

Yamashima, T., Tohma, Y., & Yamashita, J. (1992). Expression of cell adhesion molecule E-cadherin in human arachnoid villi. J. Neurosurg. 77, 749-756.

Yoshida, C., & Takeichi, M. (1982). Teratrocarcinoma cell adhesion: Identification of a cell surface protein involved in calcium-dependent cell aggregation. Cell 28, 217-224.

Yoshida-Noro, C., Suzuki, M., & Takeichi, M. (1984). Molecular nature of the calcium-dependent cell-cell adhesion system in mouse teratocarcinoma and embryonic cells studied with a monoclonal antibody. Dev. Biol. 101, 19-27.

DESMOSOMAL CADHERINS AND THEIR INTERACTIONS WITH PLAKOGLOBIN

Pamela Cowin and Sailaja Puttagunta

I. INTRODUCTION

Desmosomes are specializations of the plasma membrane that mediate strong intercellular adhesion (Cowin and Burke, 1996). Under the electron microscope, the desmosomal adhesive core appears as a

Advances in Molecular and Cell Biology,
Volume 16, pages 113-136.

ISBN: 0-7623-0143-0.

region of widened intercellular space into which rows of cross bridges project from either cell membrane. These structures overlap with each other centrally to form an electron-dense line that is a hallmark of this type of cell junction (Farquhar and Palade, 1963). In addition to their adhesive role desmosomes form the membrane attachment sites for several types of intermediate filaments that associate with the electron-dense plaques underlying the desmosomal membrane.

Mixtures of two types of transmembrane glycoproteins called desmogleins (Dsg) and desmocollins (Dsc) form the intercellular structural features of desmosomes (Buxton et al., 1993). Monovalent antibodies directed against desmocollins inhibit desmosome formation by cultured cells, providing evidence for the crucial role played by these glycoproteins in cell-cell adhesion (Cowin et al., 1984). The importance of the desmosomal adhesive role in vivo is demonstrated by a group of autoimmune diseases known collectively as pemphigus. Individuals with these disorders have antibodies directed against desmogleins and desmocollins which are deposited in epidermis and squamous tissues, causing severe blistering in these sites (Eyre and Stanley, 1987; Amagai et al., 1991; Dmochowski et al., 1993).

The amino acid sequences of both desmocollins and desmogleins display significant similarity to the ectodomains of classical cadherins but differ from them and from each other in the sequence of their cytoplasmic region (Goodwin et al., 1990; Collins et al., 1991; Mechanic et al., 1991; Nilles et al., 1991; Parker et al., 1991; Wheeler et al., 1991; Amagai et al., 1992). This review will briefly highlight the structural features and expression patterns of these proteins and will focus on their interactions with members of the plakoglobin family.

II. STRUCTURAL FEATURES OF DESMOSOMAL GLYCOPROTEINS

Desmosomal glycoproteins form two of the major subsets with the cadherin superfamily (Table 1). The best described group of proteins within this superfamily are the classical cadherins, which are calcium-dependent cell-cell adhesion and recognition proteins (Takeichi, 1991). Several members of the classical cadherin family localize to adherens junctions, which bear a superficial resemblance to the desmosome but can be distinguished from the latter by their biochemical composition and their association with the actin

Table 1

Cadherin Gene Superfamily	
CLASSICAL CADHERINS	Takeichi (1991)
DESMOGLEINS	Goodwin et al. (1990); Koch et al. (1990); Wheeler et al. (1991); Nilles et al. (1991)
DESMOCOLLINS	Mechanic et al. (1991); Collins et al. (1991); Parker et al. (1991); Koch et al. (1991)
PROTOCADHERINS	Sano et al. (1993)
FAT	Mahoney et al. (1991)
DACHSOUS	Mahoney et al. (1991)
HPT-1	Dantzig et al. (1994)
Genes with Small Regions of Cadherin-like sequence	
cRET	Iwamoto et al. (1993)
LETHAL GIANT	Klambt et al. (1989)
PTPμ	Tonks et al. (1992)

cytoskeleton (Boller et al., 1985; Hirano et al., 1987; Mege et al., 1988). Desmogleins and desmocollins are similar to classical cadherins in the sequences and arrangement of their ectodomain, which in each protein contains four consecutive extracellular repeats (EC1-4) of ~ 110-120 residues (Figure 1). The greatest similarity among classical and desmosomal cadherins is found in the stretches of negative amino acids toward either end of each repeat as well as in the prominent LDRE motifs that form binding pockets for calcium ions (Ozawa et al., 1990a; Shapiro et al., 1995). Moreover, amino acids that are involved in dimerization of classical cadherins in the plane of the membrane are highly conserved in desmosomal cadherins (Shapiro et al., 1995). Interestingly, desmocollins show more similarity to classical cadherins than to desmoglein, their neighbor in the desmosome (Figure 2; Mechanic et al., 1991).

Despite the overall similarity among desmosomal glycoproteins and classical cadherins there are subtle but important differences in the sequences of their ectodomain. Both desmosomal glycoproteins differ from each other and from classical cadherins in amino acids that play a crucial role in specifying their adhesive and recognition function (Nose et al., 1990; Shapiro et al., 1995). Furthermore, there are differences in key residues that may significantly alter the

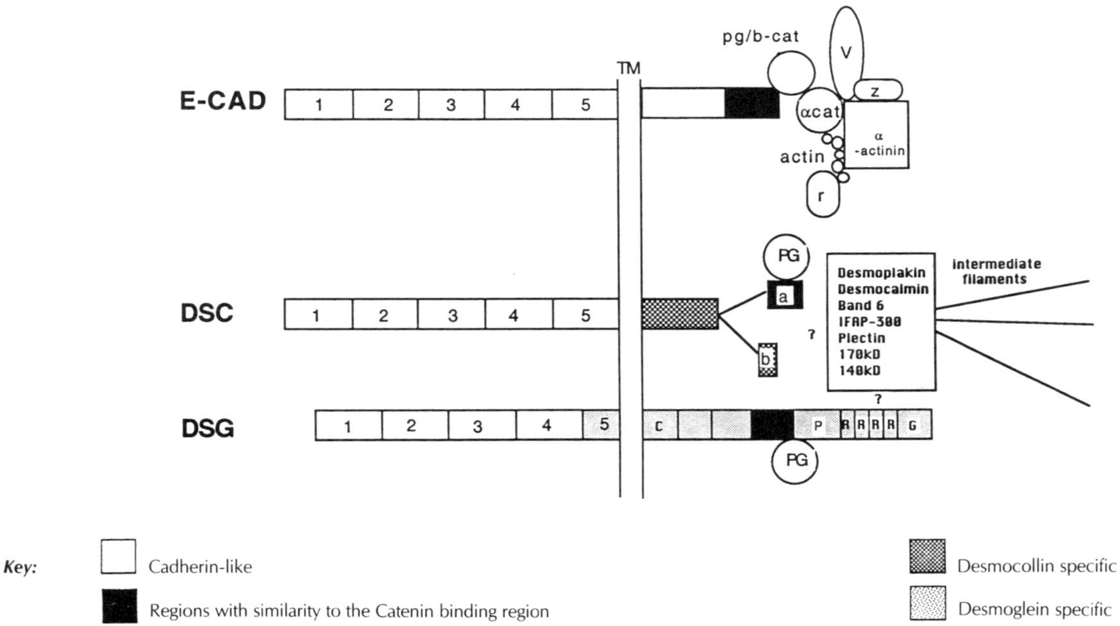

Figure 1. Diagram showing the cytoplasmic interactions of E-cadherin, desmocollins, and desmogleins, respectively. The members of the protein complex that associates with the cytoplasmic domain of E-cadherin include plakoglobin (PG), β-catenin (bcat), α-catenin (acat), vinculin (v), zyxin (z), α-actinin, actin, and radixin (r). Plakoglobin is also known to associate with the cytoplasmic domain of desmoglein. The proteins that are proposed to mediate attachment of intermediate filaments to the desmosomal glycoproteins include desmoplakin, desmocalmin, band 6, 170 kD protein, IFAP 300p, and plectin.

Numericals one to five depict the extracellular repeats of the respective proteins. TM = transmembrane domain. P = proline rich region. G = glycine rich region. C = cysteine rich region. a and b = alternatively spliced forms of desmocollins.

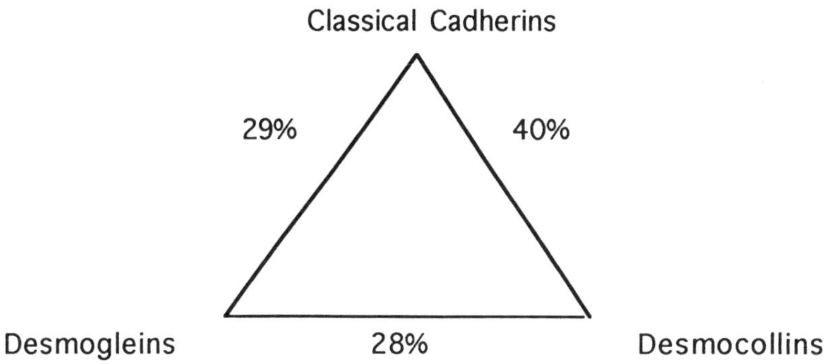

Figure 2. A schematic illustrating the degree of identity between the amino acid sequences of the ectodomain of classical cadherin, desmoglein, and desmocollin, expressed as percentages.

secondary folding of these proteins and thereby account for the distinct structural appearance of desmosomes and adherens junctions; desmosomes display considerable structural specialization in their adhesive domain whereas adherens junctions have an amorphous intercellular appearance (Goodwin et al., 1990; Mechanic et al., 1991; Cowin and Mechanic, 1994). For example, proline residues, which can break alpha-helical structure, are highly conserved among the classical cadherins but are conspicuously absent from desmogleins (Goodwin et al., 1990). Furthermore, EC1-4 of desmogleins and desmocollins contain many cysteine residues that are conserved within their subtypes but are absent from classical cadherins (Goodwin et al., 1990; Mechanic et al., 1991). These cysteines could alter intramolecular protein folding and intermolecular protein associations between desmosomal cadherins through their ability to form disulphide bridges.

Besides the four extracellular repeats, desmosomal and classical cadherins contain a fifth extracellular region, bordering the plasma membrane that is highly variable in sequence. The most noticeable feature of this region is the presence of four cysteine residues that occupy conserved positions in classical cadherins, desmocollins, and in Dsg3. These cysteines are absent, however, from EC5 of Dsg1, which is not only truncated in length but also shows allelic polymorphism (Puttagunta et al., 1994). EC5 is thought to act as a hinge to deliver the more important sites of the molecule into the intercellular space and, therefore, is able to tolerate a high degree of sequence variability. The epitopes of several adhesion-disrupting antibodies including, in Dsg1, those recognized by pemphigus foliaceus sera map to the EC5 region (Ozawa et al., 1990b; Allen et al., 1992; Geiger et al., 1992). However, antibodies may hinder the adhesive mechanism in many ways besides occupying the active site of interaction. Since clustering of cadherins into junctional sites is known to be essential for adhesive function, these antibodies may disrupt adhesion by sterically hindering lateral packing of cadherin molecules within the plane of the membrane.

The greatest divergence in sequence among desmogleins, desmocollins, and classical cadherins is found in their cytoplasmic regions (Figure 1). Desmocollins are produced as two forms by alternative splicing of the last exon (Collins et al., 1991; Parker et al., 1991). Desmogleins have a complex cytoplasmic region that is 2-3 times larger than classical cadherins (Koch et al., 1990; Nilles et al., 1991; Wheeler et al., 1991). The predominantly unique sequence features of their cytoplasmic regions account for the interaction of these three types of cadherins with different subcortical and cytoskeletal proteins (Figure 1). However, both the larger splice form of desmocollins and a central region of desmogleins contain a sequence with considerable homology to the carboxyl-terminal exon of classical cadherins. In the latter, this region is known to associate with one of a choice of three related proteins, plakoglobin, beta-catenin, or p120. These proteins regulate the adhesive function of E-cadherin, by modulating association with alpha-catenin and the actin cytoskeleton (Ozawa et al., 1989). The strong sequence similarity among desmosomal glycoproteins and classical cadherins in this domain suggests that they may share cytoplasmic interactions that are crucial to their common adhesive function.

III. BIOSYNTHESIS, ASSEMBLY, AND ADHESIVE MECHANISM

Both classical and desmosomal cadherins are produced as precursors and share a common proteolytic cleavage signal suggesting that they may be processed to their mature forms by the same enzyme. This step occurs following addition of complex sugars and prior to their exit from the medial stacks of the Golgi apparatus (Penn et al., 1987a; 1987b; Pasdar and Nelson, 1988a, 1988b, 1989; Penn et al., 1989; Pasdar et al., 1991; Shore and Nelson, 1991). The prosegment is thought to prevent adhesion from occurring in the biosynthetic compartments of the cell and its complete removal is required for the proteins to manifest their adhesive properties (Ozawa and Kemler, 1990).

Desmosomal and classical cadherins associate with several cytoplasmic proteins early during their biosynthesis. Beta-catenin and plakoglobin bind to the cytoplasmic domain of cadherin precursors while they are still in the endoplasmic reticulum. Alpha-catenin joins the E-cadherin complex after proteolytic cleavage of the prosegment has occurred and is stabilized by this association (Hirano et al., 1992; Ozawa and Kemler, 1992). Desmosomal glycoproteins are processed slightly slower than classical cadherins and take about on hour to arrive at the cell surface. Following cleavage of their prosegment, desmosomal glycoproteins enter a Triton X-100 insoluble compartment suggesting that at this stage they oligomerize or form a complex with other transmembrane components, the cytoskeleton, or elements of the sorting machinery of the cell (Pasdar and Nelson, 1989; Pasdar et al., 1991).

Junction assembly and stability is extremely sensitive to external levels of calcium (Hennings and Holbrook, 1983). Cadherins are known to bind calcium and to undergo a conformational change in the process that protects them from trypsinization (Ozawa et al., 1990a). In low calcium culture conditions that prevent cell adhesion, desmosomal cadherins arrive at the plasma membrane but turn over rapidly, with a half life of about one to two hours. In the presence of calcium and cell-contact desmosome assembly is rapid, becoming evident within 15 seconds and is complete by about two hours, and under these circumstances, desmosomal glycoproteins remain stable more than 24 hours (Hennings and Holbrook, 1983; Penn et al., 1989; Pasdar and Nelson, 1989; Pasdar et al., 1991). Classical cadherins bring about cell adhesion through homophilic interaction (Nose et al., 1988). Weak homophilic interaction has been suggested for desmogleins (Amagai et al., 1994). Although those observations cannot fully account for the known strength of adhesion mediated by desmosomes.

Several studies have indicated that phosphorylation and dephosphorylation events may control junction assembly and stability (Sheu et al., 1989). For example, kinase inhibitors such as H7 prevent the disassembly of intercellular junctions that usually occurs when extracellular calcium is removed (Citi, 1992). Other studies have shown a correlation between the level of phosphorylation of a particular junctional component and various states of cell adhesion. For example, desmocollins are reported to be more heavily phosphorylated in rapidly growing cell cultures that are actively remodeling their junctions than in confluent and quiescent cultures (Parrish et al., 1990).

It has been known for many years that cell junctions are disrupted in src transformed cells (Warren and Nelson, 1987; Volberg et al., 1991). The growing number of kinases of the src family, for example, c-Yes and C-fyn, that are enriched at adherens junctions makes it extremely likely that phosphorylation of junctional components themselves may be key targets of the action of src (Maher et al., 1985; Tsukita et al., 1991). The degree of tyrosine phosphorylation of beta catenin has been shown to correlate with the extent of impairment of adhesion of cells transformed by src. Both features were coordinately alleviated by treatment with a tyrosine kinase inhibitor and exacerbated by inhibitors of phosphatases (Matsuyoshi et al., 1992). Recently antibodies have been raised to a number of putative substrates of src and this has permitted investigation of their localization and potential function. One p125FAK, focal adhesion kinase, localizes to focal contacts where it functions, in response to ligand binding of the integrin receptor, to phosphorylate a number of cytoplasmic proteins associated with these junctions. Src, therefore, by constitutively activating p125FAK, mimics signals normally produced by cell contact and thus permits the cells to grow in an anchorage independent mode (Guan and Shalloway, 1992; Lipfert et al., 1992). A second substrate of src, called p120, has been cloned and sequenced and shown to have low but significant homology to plakoglobin, the common cytoplasmic component of intercellular adhesive junctions (Reynolds et al., 1989, 1992). p120 localizes to a subset of intercellular adherens junctions and is particularly enriched in junctions found in stratified epithelia as well as in endothelial cell junctions (Witcher and Cowin, unpublished data). p120 has been shown to associate with E-cadherin and appears to be the catenin of choice in ras transformed cells (Reynolds et al.,

1995; Kinch et al., 1995). The inability of p120 to bind alpha-catenin and thus to tether cadherins to the cytoskeleton (see below) may therefore account for the poor adhesive phenotype of such cells (Kinch et al., 1995).

IV. DIVERSITY AND EVOLUTION OF THE DESMOSOMAL CADHERINS AND THEIR ROLE IN DISEASE

More than 20 members of the classical cadherin family have been described that are differentially expressed during development and many experimental studies have established their role in regulating specific interactions between groups of cells during morphogenesis (Takeichi, 1991). Whether desmosomal glycoproteins also play a role in cell recognition is not known. However, their expression patterns even within a single tissue are complex and compatible with such a possibility. To date, studies have revealed the presence of three desmogleins and three pairs of desmocollins that show differential expression patterns in the various layers of stratified epithelia (Table 2; Koch et al., 1991a, 1991b; 1992; Arnemann et al., 1993; King et al., 1993; Theis et al., 1993; Buxton et al., 1994). Studies of epitope expression patterns indicate that many more desmogleins and desmocollins exist and it is likely that they will be as diverse as classical cadherins (Cohen et al., 1983; Cowin and Garrod, 1983; Giudice et al., 1984; Suhrbier and Garrod, 1986).

The expression patterns of desmosomal glycoproteins play a significant role in the pathology of a group of blistering diseases known as pemphigus. At least two desmogleins, Dsg1 and Dsg3, contain epitopes that are the antigenic targets of pathogenic

Table 2. Expression Patterns of Transcripts of Desmosomal Glycoproteins

Glycoprotein	*Tissue Distribution of Transcript*
Desmogleins	
Dsg1	Suprabasal layers of epidermis, tongue, and esophagus
Dsg2 (HDGC)	Basal layer of epidermis, intestinal epithelia, and MDBK cells
Dsg3 (PVA)	Lower spinous layers epidermis, tongue, and esophagus
Desmocollins	
Dsc1a, 1b	Suprabasal epidermis and tongue
Dsc 2a, 2b	Spinous layers of epidermis and all layers of tongue
Dsc 3a, 3b	Basal layer of epidermis, exocervix, and esophagus

antibodies circulating in the sera of patients with pemphigus. Patients with pemphigus vulgaris have deep blistering of squamous epithelia principally involving the mouth and tongue with the split occurring just above the basal cells. Dsg3, the antigenic target of these antibodies, is found in the lower layers of epidermis, tongue, and mucosa (Amagai et al., 1991, 1992). In contrast, patients with the rarer form of blistering disease known as pemphigus foliaceus who have antibodies directed against Dsg1, show blistering of the superficial layers of epidermis, where Dsg1 is maximally expressed (Eyre and Stanley, 1987).

Several human desmosomal and classical cadherin genes, DSG1, 2, and 3, DSC1, and N-CAD have been mapped to chromosome 18 (Buxton et al., 1993). In mice, a close linkage has been reported between DSC3 and DSG1 (Buxton et al., 1994). A comparison of the gene structures of desmosomal and classical cadherins supports the possibility that they have evolved from a common ancestral gene (Puttagunta et al., 1994). The genes for classical cadherins and DSG1 show an identical and distinctive pattern of intron/exon boundaries that split consecutive repeats at different positions. This arrangement bears no obvious relationship to the functional domains of the protein, yet its conservation implies that it serves a crucial role (Puttagunta et al., 1994). By interrupting consecutive repeats in different places, introns decrease the length of homologous sequences, thereby reducing the potential for misalignment of sister chromatids and consequent unequal recombination (Klein, 1988). In this manner, the evolution of cadherins with differing numbers of extracellular repeats is prevented.

It would appear that the ancestral gene, comprising a single repeat unit, duplicated twice to give rise to a four repeat sequence. The modern form of exon boundaries were then established before further divergence produced the desmogleins, desmocollins, and classical cadherins. The variable number of repeats seen in other members of the cadherin superfamily, such as the protocadherins which have between six and seven repeats and Drosophila cadherins such as fat which has 34 repeats, suggests that these forms may have evolved by unequal recombination prior to the introduction of introns (Mahoney et al., 1991; Sano et al., 1993).

Evidence of intergenic recombination and gene conversion can be seen both among desmocollins and desmogleins. For example, the allelic polymorphism reported in bovine Dsg1 is produced either by double-reciprocal crossover or gene conversion of the fifth region by a related gene (Puttagunta et al., 1994). Similarly, a close inspection of desmocollin sequences reveals evidence of recombination between type 1 and type 2 in several regions (Cowin and Mechanic, 1994). Clearly, such events have contributed to the diversification of the cadherin superfamily, resulting in a large repertoire of proteins with the complexity and subtlety required for a role in cellular recognition.

V. CYTOPLASMIC INTERACTIONS

In many cell types, expression of exogenous classical cadherins leads to significant changes in the morphology and adhesive behavior of the host cell. For example, PC12 neuro-pheochromocytoma cells, which remain rounded when seeded on normal 3T3 cells, will extend neurites when plated upon fibroblasts expressing N-cadherin. This change in morphology is dependent upon functional calcium channels and a cholera toxin-sensitive G-protein and can be reproduced by direct activation of these downstream effectors (Saffell et al., 1992). Another example is found when E-cadherin is transfected into L-cells. Fibroblasts expressing E-cadherin change their adhesive behavior and morphology to one reminiscent of cultured epithelial cells (Nose et al., 1988). These changes are accompanied by significant rearrangements in the subcortical and filamentous cytoskeleton (McNeil et al., 1990). Similar studies with mutant cadherins showed these effects of cadherin expression to be dependent upon the presence of the last exon of their cytoplasmic domain. Mutants deleted in this region display normal binding properties of the ectodomain but fail to cluster into junctional sites, to associate with and reorganize the cytoskeleton, and were unable to support stable intercellular adhesion (Nagafuchi and Takeichi, 1988; Ozawa et al., 1989). This region is important for the binding of the catenin complex of cytoplasmic regulatory proteins (Ozawa et al., 1989). Overexpression of the transmembrane cytoplasmic domains produces a dominant negative effect on cell adhesion through their ability to compete with endogenous cadherins for the binding of these proteins (Kintner, 1992). These results suggest that the cadherin cytoplasmic sequences or the proteins with which they interact modulate the ability of the ectodomain to bring about cell adhesion by eliciting a signaling cascade and/or by tethering cadherins to the cytoskeleton and clustering them into junctional sites.

Expression of the desmosomal glycoprotein cytoplasmic regions as chimeric proteins attached to the transmembrane region of the gap junction protein connexin suggests they too exert an essential regulatory role on junction assembly that can operate in the absence of the ectodomain. For example, expression of connexin chimeras containing the last exon of Dsc1a, which has homology to the catenin-binding region of classical cadherins, causes recruitment of plaque components and intermediate filaments to the plasma membrane (Troyanovsky et al., 1993). In contrast, expression of the desmoglein cytoplasmic domain as a connexin chimera acts as a dominant negative mutant inhibiting assembly of endogenous desmosomal proteins into cell junctions (Troyanovsky et al., 1993).

These observations focused attention on the cytoplasmic interactions of each set of cadherins and upon their mode of linkage to the cytoskeleton. A plethora of proteins have been localized to the submembranous densities of adherens junctions as well as to the cytoplasmic plaques of desmosomes (Figure 1). In the adherens junction linkage has been established between vinculin, alpha-actinin, and F-actin (Tsukita and Tsukita, 1993). Similarly, interaction between zyxin and alpha actinin and between radixin and the barbed end of actin has been demonstrated (Tsukita and Tsukita, 1989; Crawford et al., 1992). The linkage between classical cadherins and this assemblage of proteins is mediated by the catenins. The carboxyl terminus of cadherins bind to the central repeats of plakoglobin or beta catenin (Ozawa and Kemler, 1992). These proteins in turn bind through their N-terminal regions to alpha-catenin. The latter is thought to bind directly and indirectly (through alpha-actinin) to F-actin (Cowin and Burke, 1996). This complex of proteins provides a vertical linkage between cadherins and the actin cytoskeleton and leads to clustering of cadherins in the plane of the membrane.

Presently, five proteins have been proposed to bind intermediate filaments to the desmosomal plaque. For two of these, plectin and IFAP-300, this assertion rests on their ability to bind to intermediate filaments in vitro and their localization at both desmosomes and hemidesmosomes, which are the major sites of membrane anchorage for these filaments (Wiche et al., 1984; Skalli et al., 1994). Desmocalmin, a 22OkD calmodulin-binding constituent of subcellular fractions enriched in epidermal desmosomes, has been found to cosediment with intermediate filaments (Tsukita and Tsukita, 1985). Band 6, another major component of these biochemical preparations, has been shown to bind to simple epithelial keratins in blot overlay assays (Kapprell et al., 1988; Hatzfeld et al., 1994). Desmoplakins 1 and 11, which are major constituents of all desmosomal plaques, share a region of sequence similarity with the 1b rod domain of intermediate filaments (Green et al., 1990). This sequence is also found in plectin, a known intermediate filament bundling protein, as well as in bullous pemphigoid antigen, a plaque component of hemidesmosomes, and has been proposed to facilitate binding of intermediate filaments to these proteins (Green et al., 1990). Overexpression of the carboxyl-terminal half of desmoplakins that contains this sequence supports this possibility. This mutant protein coaligns with the intermediate filaments and causes them to collapse away from the membrane (Stappenbeck and Green, 1992). The C-terminal domain of desmoplakins binds in blot overlay assays to the N-termini of basic epidermal keratins (Kouklis et al., 1995). The N-terminal domain of desmoplakins is required for its association with desmosomes possibly through binding to desmocollins.

Of the myriad of proteins that localize to the desmosomes and adherens junctions only one, a component called plakoglobin, is shared by both (Cowin et al., 1986). Its presence at the two major types of intercellular adhesive junction suggests that it may function to coordinately regulate the adhesive properties of the cell. Plakoglobin was first described as a component of desmosome preparations (band 5) and was initially immunolocalized to these junctions (Cowin et al., 1985; Gorbsky et al., 1985). Subsequently, it was found in a wide variety of cell-cell adhesions junctions and in considerable amounts in the cytoplasm (Cowin et al., 1986; Franke et al., 1987; Kapprell et al., 1987). This pattern of localization suggests that plakoglobin is not involved in filament association as desmosomes and adherens junctions bind different elements of the cytoskeleton. Plakoglobin is absent from cell-substratum junctions such as focal contacts and hemidesmosomes, which share some plaque proteins and engage the same filaments as their cell-cell junction counterparts but employ integrins as their adhesive component. This feature of -plakoglobin localization strongly suggests that its role is closely tied to cadherin function. Plakoglobin has been coprecipitated as part of complexes associated with both desmosomal and classical cadherins (Korman et al., 1989; Knudsen and Wheelock, 1992; Peifer et al., 1992; Piepenhagen and Nelson, 1993). Plakoglobin binds to the central region of the cytoplasmic sequence of Dsg1 that shares homology with the catenin-binding domain of classical cadherins (Mathur et al., 1994). Interestingly, the dominant negative effect produced by overexpression of the desmoglein cytoplasmic domain is relieved when the plakoglobin-binding region is deleted from the construct (Shafer et al., 1993). This suggests that the mutant may prevent junction assembly by binding and depleting the cellular pool of plakoglobin that is necessary for the function of endogenous desmogleins. Plakoglobin also binds to the larger alternatively-spliced form of desmocollin (Witcher et al., 1996). Desmoglein bind to the first three repeats of the central region of plakoglobin whereas desmocollin appears to bind to both ends of this domain (Witcher et al., 1996).

Strong sequence similarity is found among plakoglobin, beta catenin, and the product of a *Drosophila* segment polarity gene, *Armadillo* (Table 3; Franke et al., 1989; Peifer and Weischaus, 1990; McCrea et al., 1991). Segment polarity genes are thought to act in a pathway that transmits positional information across the body segment of *Drosophila* embryos, with their action first becoming

Table 3. "ARM" Family Proteins

>60% Amino Acid Sequence Identity to *Armadillo*
Plakoglobin
Beta Catenin
Armadillo
>50% Identity to "ARM" Consensus Sequence
Band 6
P120
APC
smgGDS
SRP-1/Importin/RCH1/Pendulin
PP2A
Elongation Factor 3
Adaptins

evident in the morphology of the cuticle overlying the epidermis (Ingham, 1991; Nusse and Varmus, 1992). *Armadillo* is required for transmission of the *wingless* signal (Riggleman et al., 1990). The *wingless* gene encodes a secreted cysteine rich protein that binds to the extracellular matrix and membranes of neighboring cells through an unknown receptor (Nusse and Varmus, 1992). Expression of *wingless* leads to postranscriptional up-regulation of the expression of a cytoplasmic pool of Armadillo protein in neighboring cells (Riggleman et al., 1990; Peifer et al., 1994b).

We have shown that PC12 pheochromocytoma cells expressing Wnt-1, the human homologue of Wingless, show a post-transcriptional elevation in cytosolic and membrane bound forms of plakoglobin. This is accompanied by a significant alteration in the cellular morphology and adhesive behavior giving rise to an epithelial appearance (Bradley et al., 1993). Increased stability of the beta-catenin-E-cadherin complex accompanied by an increase in cell adhesion has also been found in response to Wnt-1 expression by AtT20 cells (Hinck et al., 1994). These studies suggest demonstrate that the mechanisms of patterning by *wingless* pathway in *Drosophila* have been highly conserved in vertebrates and that plakoglobin and beta-catenin are necessary components in the reception or implementation of *wnt* signals. Further support for this comes from studies in *Xenopus*. Overexpression of *wnt* plakoglobin, or beta-catenin induces duplication of the embryonic axis in this organism. This phenotype correlates with both cytosolic and nuclear localization of plakoglobin or beta-catenin suggesting that these proteins may transmit patterning signals directly to nuclear targets (Karnovsky, 1995; Funayama, 1995). Consistent with this notion, expression of classical or desmosomal cadherins which retain plakoglobin and beta-catenin at the membrane reverse or oppose these phenotypic changes (Karnovsky, 1995; Heasman, 1994).

A reexamination of the role of the cytosolic pools of plakoglobin and beta-catenin has been provoked by the finding of a cytosolic complex between these proteins and APC, the product of the tumor supressor gene (Rubinfeld et al., 1993; Su et al., 1993). APC is linked to an inherited disease of the colon, familial adenopolyposis. The product of this gene has low but significant homology to plakoglobin, and mutations that result in its truncation have been implicated in a number of sporadic colon carcinomas (Kinsler et al., 1991; Nishisho et al., 1991). APC has been shown to target plakoglobin and beta-catenin for degradation and to arrest cells in Go (Polakis 1995; Baeg et al., 1995). Mutant APC is deficient in both these functions. Thus, a correlation exists between an ability to lower cytosolic levels of plakoglobin and beta-catenin and growth arrest. Since cadherins also titrate these proteins from the cytosol to the membrane it is possible that this change in the equilibrium could porovide the mechanism for contact-inhibition of cell growth.

In summary, desmosomal glycoproteins form two distinct subsets of the cadherin superfamily. While they clearly share with the classical cadherins a role in cell adhesion, it remains to be established whether they contribute to other functions associated with these proteins such as cell recognition and induction of cell differentiation. The cytoplasmic interactions of the desmosomal glycoproteins are beginning to yield to a combination of molecular and genetic analysis. Several plakoglobin-like proteins have important signaling roles functions outside of cell junctions. Deciphering the mechanism by which these proteins participate in development and growth regulatory pathways promises to be an exciting chapter in the investigation of the role of junction-associated proteins.

ACKNOWLEDGMENTS

Work in our laboratory is supported by the National Institutes of Health GM47429 and the Pew Charitable Trusts.

REFERENCES

Allen, E.M., Guidice, G., & Diaz, L. (1992). Production of human DG1 fusion proteins and their reactivity with PF autoantibodies. J. Invest. Dermatol. 98, 590.

Amagia, M., Klaus, K.V., & Stanley, J.R. (1991). Autoantibodies against a novel epithelial cadherin in pemphigus vulgaris, a disease of cell adehsion. Cell 67, 869-877.

Amagai, M., Karpati, S., Prussick, R., Klasu-Kovtun, V., & Stanley, J.R. (1992). Autoantibodies against the aminoterminal cadherin-like binding domain of pemphigus vulgaris antigen are pathogenic. J. Clin Invest. 90, 919-1926.

Amagai, M., Karpati, S., Klasu-Kovtun, V., Udev, M.C., & Stanley, J.R. (1994). Extracellular domain of pemphigus vulgaris antigen (desmglein 3) mediates weak homophilic adhesion. J. Invest. Dermatol. 102, 402-407.

Arnemann, J., Sullivan, K.H., Magee, A.I., King, I.A., & Buxton, R.S. (1993). Stratification-related expression of isoforms of the desmosomal cadherins in human epidermis. Journal of Cell Science 104, 741-750.

Boller, K., Vestweber, D., & Kemler, R. (1985). Cell adhesion molecule uvomorulin is localised in the intermediate junctions of adult intestinal epithelial cells. J. Cell Viol. 100, 327-332.

Bradley, R.S., Cowin, P., & Brown, A.M.C. (1993). Expression of Wnt-1 in PC12 cells results in modulation of plakoglobin and E-cadherin and increased cellular adhesion. J. Cell Biol. 123, 1857-1865.

Buxton, R.S., Cowin, P., Franke, W.W., Garrod, D.R., Green, K.J., King, I.A., Koch, P.J., Magee, A.I., Rees, D.A., Stanley, J.R., & Steinberg, M.S. (1993). Nomenclature of the desmosomal cadherins. J. Cell Biol. 121, 481-483.

Buxton, R.S., Wheeler, G.N., Pidsley, S.C., Marsden, M.D., Adans, M.J., Jenkins, N.A., Gilbert, D.J., & Copeland, N.G. (1994). Mouse desmocollin DSC3 and desmoglein DSG1 genes are closely linked in the proximal region of chromosome 18. Genomics.

Citi, S. (1992). Protein kinase inhibitors prevent junction dissociation induced by low extracellular calcium in MDCK cells. J. Cell Biol. 117, 169-178.

Cohen, S.M., Gorbsky, G., & Steinberg, M.S. (1983). Immunochemical characterization of related families of glycoproteins in desmosomes. J. Biol. Chem. 258, 2621-2627.

Collins, J.E., Legan, P.K., Kenny, T.P., MacGarvie, J., Holton, J.L., & Garrod, D.R. (1991). Cloning and sequence analysis of desmosomal glycoproteins 2 and 3 (desmocollins): Cadherin-like desmosomal adhesion molecules with heterogeneous cytoplasmic domains. J. Cell Biol. 113, 381-391.

Cowin, P., & Burke, B. (1996). Cytoskeleton-membrane interactions. Curr. Opin. Cell Biol. 8, 56-65.

Cowin, P., & Garrod, D.R. (1983). Antibodies to epithelial desmosomes show wide tissue and species cross-reactivity. Nature 302, 148-150.

Cowin, P., Kapprell, H.-P., Franke, W.W., Tamkun, J., & Hynes, R.O. (1986). Plakoglobin: A protein common to different kinds of intercellular adhering junctions. Cell 46, 1063-1073.

Cowin, P., Kapprell, P.H.P., & Franke, W.W. (1985). The complement of desmosomal plaque proteins in different cell types. J. Cell Biol. 101, 1442-54.

Cowin, P., Mattey, D., & Garrod, D.R. (1984). Identification of desmosomal surface components (desmocollins) and inhibition of desmosome formation by specific FAB'. J. Cell Sci. 70, 41-60.

Cowin, P., & Mechanic, S. (1994). Desmosomal cadherins and their cytoplasmic Interactions. In: *Molecular Mechanism of Epithelial Cell Junctions: From Development to Disease.* R.G. Landes, Austin, TX, pp. 141-157.

Crawford, A.W., Michelson, J.W., & Beckerle, M. (1992). An interaction between zyxin and alpha-actinin. J. Cell Biol. 116, 1381-1393.

Dantzig, A.H., Hoskins, J., Tabas, L.B., Bright, S., Shepard, R.L., Jenkins, I.L., Duckworth, D.C., Sportsman, J.R., Mackinsen, D., Rosteck, P.R., & Skatrud, P.L. (1994). Association of Intestinal peptide transport with a protein related to the cadherin superfamily. Science 264, 430-433.

Dmochowski, M., Hashimoto, T., Garrod, D.R., & Nishikawa, T. (1993). Desmocollins I and II are recognized by certain sera from patients with various types of pemphigus, particularly brazilian pemphigus foliaceus. J. Invest. Dermatol. 100, 380-384.

Eyre, R.W., & Stanley, J.R. (1987). Human autoantibodies against a desmosomal protein complex with a calcium-sensitive epitope are characteristics of pemphigus foliaceus patients. J. Exp. Med. 165, 1719-1724.

Farquhar, M.G., & Palade, G.E. (1963). Junctional complexes in various epithelia. J. Cell Biol. 17, 375-412.

Franke, W.W., Goldschmidt, M.D., Zimblemann, R., Mueller, H.M., Schiller, D.L., & Cowin, P. (1989). Molecular cloning and amino acid sequence of plakoglobin, common junctional plaque protein. Proc. Natl. Acad. Sci. USA 86, 4027-4031.

Franke, W.W., Kapprell, H.-P., & Cowin, P. (1987). Immunolocalization of plakoglobin in endothelial junctions: Identification as a special type of *Zonulae Adherentes*. Biol. Cell 59, 205-218.

Geiger, B., Volberg, T., Sabanay, I., & Volk, T. (1992). A-Cam: An adherens junction-specific cell adhesion molecule. In: *The Cell in Contact*. John Wiley, New York, pp. 55-79.

Giudice, G.J., Cohen, S.M., Patel, N.H., & Steinberg, M.S. (1984). Immunological comparison of desmosomal components from several bovine tissues. J. Cell Biochem. 26, 35-45.

Goodwin, L., Hill, J.E., Raynor, K., Raszi, L., Manabe, M., & Cowin, P. (1990). Desmoglein shows extensive homology to the cadherin family of cell adhesion molecules. Biochem. Biophys. Res. Com. 173, 1224-30.

Gorbsky, G., Cohen, S.M., Shida, H., Giudice, G.J., & Steinberg, M.S. (1985). Isolation of the non-glycosylated proteins of desmosomes and immunolocalization of a third plaque protein: desmoplakin III. Proc. Natl. Acad. Sci. USA 82, 810-4.

Green, K.J., Parry, D.A., Steinert, P.M., Virata, M.L., Wagner, R.M., Angst, B.D., & Nilles, L.A. (1990). Structure of the human desmoplakins. Implications for function in the desmosomal plaque. J Biol Chem 265, 2603-12.

Guan, J.-L., & Shalloway, D. (1992). Regulation of focal adhesion-associated protein tyrosine kinase by both cellular adhesion and oncogenic transformation. Nature 358, 690-692.

Hennings, H., & Holbrook, K. (1983). Calcium Regulation of Cell-Cell Contact and Differentiation of Epidermal Cells in Culture. Exp. Cell Res.. 143, 127-142.

Herrenknecht, K., Ozawa, M., Eckerskorn, C., Lottspeich, F., Lenter, M., & Kemler, R. (1991). The uvomorulin-anchorage protein alpha catenin is a vinculin homologue. Proc. Natl. Acad. Sci. 88, 9156-60.

Hinck, L., Nelson, W.J., & Papkoff, J. (1994). Wnt-1 modulates cell-cell adhesion in mammalian cells by stabilizing β-catenin binding to the cell adhesion protein cadherin. J. Cell Biol. 124, 729-741.

Hirano, S., Kimoto, N., Shimoyama, Y., Hirohashi, S., & Takeichi, M. (1992). Identification of a neural alpha-catenin as a key regulator of cadherin function and multicellular organization. Cell 70, 293-301.

Hirano, S., Nose, A., Hatta, K., Kawakami, A., & Takeichi, M. (1987). Calcium-dependent cell-cell adhesion molecules (cadherins): Subclass specificities and possible involvement of actin bundles. J. Cell Biol. 105, 2501-2510.

Ingham, P. (1991). Segment polarity genes and cell patterning within the *Drosophila* body segment. Curr. Biol. 1, 261-267.

Iwamoto, T., TTaniguchi, M., Asai, N., Ohkusu, K., & Nakashima, I. (1993). cDNA Cloning of mouse ret proto-oncogene and its sequences similarity to the cadherin family. Oncogene 8, 1086-1091.

Kapprell, H.-P., Cowin, P., & Franke, W.W. (1987). Biochemical characterization of the soluble form of the junctional plaque protein, plakoglobin, from different cell types. Eur. J. Biochem. 166, 505-517.

Kapprell, H.P., Owaaribe, K., & Franke, W.W. (1988). Identification of a basic protein of Mr 75,000 as an accessory desmosomal plaque protein in stratified and complex epithelia. J. Cell Biol. 106, 1679-91.

King, I.A., Tabiowo, A., Purkis, P., Leigh, I., & Magee, A.I. (1993). Expression of distinct desmocollin isoforms in human epidermis. J. Invest. Dermatol. 100, 373-379.

Kinsler, K., Nilbert, M.C., Su, L., Vogelstein, B., Bryan, T.M., Levy, D.B., Smith, K.J., Presinger, A.C., Hedge, P., McKechnie, D., Finniear, Markham, A., Groffen, J., Boguski, M.S., Altshol, S.F., Mirii, A., Ando, M., Miyoshi, Y., Miki, Y., Nishisho, I., & Nakamura, Y. (1991). Identification of FAP locus genes from chromosome 5q21. Science 253, 661-669.

Kintner, C. (1992). Regulation of embryonic cell adhesion by the cadherin cytoplasmic domain. Cell 69, 225-236.

Klambt, C., Muller, S., Lutzelschwab, R., Rossa, R., Totzke, F., & Schmidt, O. (1989). The *Drosophila* melanogaster l(2)gl gene encodes a protein homologous to the cadherin cell-adhesion molecule family. Dev Biol 133, 425-36.

Klein, H. (1988). Recombination between repeated yeast genes. In: *The Recombination of Genetic Material.* Academic Press, New York, pp. 384-421.

Knudsen, K.A., & Wheelock, M.J. (1992). Plakoglobin, or an 83-kD homologue distinct from beta-catenin, interacts with E-cadherin and N-cadherin. J. Cell Biol. 118, 671-679.

Koch, P., Goldschmidt, M., Walsh, M.J., Zimbleman, R., Schmelz, M., & Franke, W. (1991a). Amino acid sequence of bovine muzzle epithelial desmocollin derived from cloned cDNA: A novel subtype of the desmosomal cadherins. Differentiation 47, 29-36.

Koch, P., Goldschmidt, M.D., Walsh, M.J., Zimbleman, R., & Franke, W.W. (1991b). Complete amino acid sequence of the epidermal desmoglein precursor polypeptide and identification of a second type of desmoglein gene. Eur J. Cell Biol. 55, 200-208.

Koch, P.J., Goldschmidt, M.D., Zimbelmann, R., Troyanovsky, R., & Franke, W.W. (1992). Complexity and expression patterns of the desmosomal cadherins. Proc. Natl. Acad. Sci. USA 89, 353-357.

Koch, P.J., Walsh, M.J., Schmelz, M., Goldschmidt, M.D., Zimbelmann, R., & Franke, W.W. (1990). Identification of desmoglein a constititive desmosomal glycoprotein as a member of the cadherin family of cell adhesion molecules. Eur. J. Cell Biol. 53, 1-12.

Korman, N.J., Eyre, R.W., Klaus-Kovtun, V., & Stanley, J.R. (1989). Demonstration of an adhering junction molecule (plakoglobin) in the autoantigens of pemphigus foliaceus and pemphigus vulgaris. N. Engl. J. Med. 321, 641-635.

Lipfert, L., Haimovich, B., Schaller, M.D., Cobb, B.S., Parsons, J.T., & Brugge, J.S. (1992). Integrin dependent phosphorylation and activation of the protein tyrosine kinase pp125 FAK in platelets. J. Cell Biol. 119, 905.

Maher, P., Paasquale, E.B., Wang, J.Y., & Singer, S.J. (1985). Phosphotyrosine-containing proteins are concentrated in focal adhesions and intercellular junctions in normal cells. Proc. Natl. Acad. Sci. 82, 6576-6580.

Mahoney, P.A., Weber, U., Onofrechuk, P., Biessmann, H., Bryant, P.J., & Goodman, C.S. (1991). The fat tumor suppressor gene in Drosophila encodes a novel member of the cadherin gene superfamily. Cell 67, 853-68.

Mathur, M., Goodwin, L., & Cowin, P. (1994). The interactions of a desmosomal cadherin, Dsg1, with plakoglobin. Journal of Biochemistry 269, 1-6.

Matsuyoshi, N., Hamaguchi, M., Taniguchi, S., Nagafuchi, A., Tsukita, S., & Takeichi, M. (1992). Cadherin-mediated cell-cell adhesion is perturbed by v-src tyrosine phosphorylation in metastatic fibroblasts. J. Cell Biol. 118, 703-714.

McCrea, P.D., Brieher, W.M., & Gumbiner, B.M. (1993). Induction of a secondary body axis in *Xenopus* by antibodies to β-catenin. J. of Cell Biol. 123, 477-484.

McCrea, P.D., Turck, C.W., & Gumbiner, B. (1991). A homolog of the armadillo protein in *Drosophila* (plakoglobin) associated with E-cadherin. Science 254, 1359-61.

McNeil, H., Ozawa, M., Kemler, R., & Nelson, W.J. (1990). Novel function of the cell adhesion molecule uvomorulin as an inducer of cell surface polarity. Cell 62, 309-316.

Mechanic, S., Raynor, K., Hill, J.E., & Cowin, P. (1991). Desmocollins form a novel subset of the cadherin family of cell adhesion molecules. Proc. Natl. Acad. Sci. USA 88, 4476-4480.

Mege, R., Matsuzaki, F., Gallin, W.J., Goldberg, J.I., Cunningham, B.A., & Edelman, G.M. (1988). Construction of epithelioid sheets by transfection of mouse sarcoma cells with cDNAs for chicken cell adhesion molecules. Proc. Natl. Acad. Sci. USA 85, 7274-8.

Nagafuchi, A., & Takeichi, M. (1988). Cell binding function of E-cadherin is regulated by the cytoplasmic domain. EMBO J 7, 3679-3684.

Nagafuchi, A., Takeichi, M., & Tsukita, S. (1991). The 102 kd cadherin-associated protein: Similarity to vinculin and posttranscriptional regulation of expression. Cell 65, 849-57.

Nilles, L.A., Parry, D.A., Powers, E.E., Angst, B.D., Wagner, R.M., & Green, K.J. (1991). Structural analysis and expression of human desmoglein: A cadherin-like component of the desmosome. Journal of Cell Science 99, 809-821.

Nishisho, I., Nakamura, Y., Miyoshi, Y., Miki, Y., Ando, H., Hirii, A., Koyama, K., Utsunomiya, J., Baba, S., Hedge, P., Markham, A., Krush, A., Peterson, G., Hamilton, S., Nilbert, M.C., Levy, D.B., Bryan, T.M., Presinger, A.C., Smith, K.J., Su, L., Kinsler, K.W., & Vogelstein, B. (1991). Mutations of Chromosome 5q21 Genes in FAP and Colorectal Cancer Patients. Science 253, 665-669.

Nose, A., Nagafuchi, A., & Takeichi, M. (1988). Expressed recombinant cadherins mediate cell sorting in model systems. Cell 54, 993-1001.

Nose, A., Tsuji, K., & Takeichi, M. (1990). Localization of Specificity Determining Sites in Cadherins Cell Adhesion Molecules. Cell 61, 147-155.

Nusse, R., & VVarmus, H. (1992). Wnt genes. Cell 69, 1073-1087.

Oda, H., Uemura, K., Shiomi, K., Nagafuchi, A., Tsukita, S., & Takeichi, M. (1993). Identification of a Drosophila homolog of alpha-catenin and its association with the armadillo protein. J. Cell Biol. 121, 1133-1141.

Olson, D.J., Christian, J.L., & Moon, R.T. (1993). Effect of Wnt-1 and Related Proteins on Gap Junctional Communication in *Xenopus* Embryos. Science 252, 1173-1176.

Ozawa, M., Baribault, H., & Kemler, R. (1989). The cytoplasmic domain of the cell adhesion molecule uvomorulin associates with three independent proteins structurally related in different species. EMBO J. 8, 1711-1717.

Ozawa, M., Engel, J., & Kemler, R. (1990a). Single amino acid substitutions in one calcium binding site of uvomorulin abolish the adhesive function. Cell 63, 1033-1038.

Ozawa, M., Hoschutzky, H., Herrenknecht, K., & Kemler, R. (1990b). A possible new adhesive site in the cell-adhesion molecule uvomorulin. Mechanisms of Development 33, 49-56.

Ozawa, M., & Kemler, R. (1990). Correct proteolytic cleavage is required for the cell adhesive function of uvomorulin. J. Cell Biol. 111, 1645-1650.

Ozawa, M., & Kemler, R. (1992). Molecular organization of the uvomorulin-catenin complex. J. Cell Biol. 116, 989-96.

Parker, A.E., Wheeler, G.N., Arnemann, J., Pidsley, S.C., Ataliotis, P., Thomas, C.L., Rees, D.A., Magee, A.I., & Buxton, R.S. (1991). Desmosomal glycoproteins II and III. Cadherin-like junctional molecules generated by alternative splicing. J. Biol. Chem. 266, 10438-10445.

Parrish, E.P., Marston, J.E., Mattey, D.L., Measures, H.R., Venning, R., & Garrod, D.R. (1990). Size heterogeneity, phosphorylation, and transmembrane organisation of desmosomal glycoproteins 2 and 3 (desmocollins) in MDCK cells. J. Cell Sci. 96, 239-248.

Pasdar, M., Krzeminski, K.A., & Nelson, W.J. (1991). Regulation of desmosome assembly in MDCK cells: Coordination of membrane core and cytoplasmic plaque domain assembly at the plasma membrane. J. Cell Biol. 113, 645-657.

Pasdar, M., Li, Z., & Chan, H. (1993). The role of phosphorylation in assembly and disassembly of desmosomes in MDCK epithelial cells. Molecular Biology of the Cell 4, 438a.

Pasdar, M., & Nelson, W.J. (1988a). Kinetics of desmosome assembly in Madin-Darby canine kidney epithelial cells: Temporal and spatial regulation of desmoplakin organization and stabilization upon cell-cell contact. I. Biochemical analysis. J. Cell Biol. 106, 677-85.

Pasdar, M., & Nelson, W.J. (1988b). Kinetics of desmosome assembly in Madin-Darby canine kidney epithelial cells: Temporal and spatial regulation of desmoplakin organization and stabilization upon cell-cell contact. II. Morphological analysis. J. Cell Biol. 106, 687-95.

Pasdar, M., & Nelson, W.J. (1989). Regulation of desmosome assembly in epithelial cells: Kinetics of synthesis, transport, and stabilization of desmoglein I, a major protein of the membrane core domain. J. Cell Biol. 109, 163-77.

Peifer, M., Berg, S., & Reynolds, A. (1994a). A repeating amino acid motif defines a new relationship among proteins with roles in signaling, tumor suppression, cell-cell adhesion, and guanine nucleotide exchange. Cell 76, 789-791.

Peifer, M., McCrea, P., Green, K.J., Wieschaus, E., & Gumbiner, B.M. (1992). The vertebrate Adhesive Junction Proteins beta-catenin and plakoglobin and the Drosophila segment Polarity gene armadillo form a multigene family with similar properties. J. Cell Biol. 118, 681-691.

Peifer, M., Sweeton, D., Cassey, M., & Wieschaus, E. (1994b). Wingless signal and Zeste-white 3 kinase trigger opposing changes in the intracellular distribution of armadillo. Development 120, 369-380.

Peifer, M., & Weischaus, E. (1990). The segment polarity gene *armadillo* encodes a functionally modular protein that is the *Drosophila* homolog of human plakoglobin. Cell 63, 1167-1178.

Penn, E.J., Burdett, I.D., Hobson, C., Magee, A.I., & Rees, D.A. (1987a). Structure and assembly of desmosome junctions: Biosynthesis and turnover of the major desmosome components of Madin-Darby canine kidney cells in low calcium medium. J. Cell Biol. 105, 2327-34.

Penn, E.J., Hobson, C., Rees, D.A., & Magee, A.I. (1987b). Structure and assembly of desmosome junctions: Biosynthesis, processing, and transport of the major protein and glycoprotein components in cultured epithelial cells. J. Cell Biol. 105, 57-68.

Penn, E.J., Hobson, C., Rees, D.A., & Magee, A.I. (1989). The assembly of the major desmosome glycoproteins of Madin-Darby canine kidney cells. Febs Lett. 247, 13-6.

Piepenhagen, P.A., & Nelson, W.J. (1993). Defining E-cadherin-associated protein complexes in epithelial cells: Plakoglobin, Beta, and gamma catenin are distinct components. Journal of Cell Science 104, 751-762.

Puttagunta, S., Mathur, M., & Cowin, P. (1994). Structure of DSG1, the bovine desmosomal cadherin gene encoding the pemphigus foliaceus antigen. J. Biol. Chem. 269, 1949-1955.

Reynolds, A.B., Herbert, L., Cleveland, J.L., Berg, S.T., & Gaut, J.R. (1992). p120, a novel substrate of protein tyrosine kinase receptors and of p60 vsrc, is related to cadherin-binding factors Beta catenin, plakoglobin and armadillo. Oncogene 7, 2439-2445.

Reynolds, A.B., Rosel, D.J., Kanner, S.B., & Parsons, J.T. (1989). Transformation-specific tyrosine phosphorylation of a novel cellular protein in chicken cells expressing oncogenic variants of the avian cellular *src* gene. Mol Cell Biol. 9, 629-638.

Riggleman, R., Schedl, P., & Wieschaus, E. (1990). Spatial expression of the Drosophila segment polarity gene *armadillo* is post-transcriptionally regulated by *wingless*. Cell 63, 549-560.

Rubinfeld, B., Souza, B., Albert, I., Muller, O., Chamberlain, S., Masiarz, F.R., Munemitsu, S., & Polakis, P. (1993). Association of the APC gene product with beta cateninn. Science 262, 1731-1734.

Saffell, J.L., Walsh, F., & Doherty, P. (1992). Direct activation of second messenger pathways mimics cell adhesion molecule-dependent neurite outgrowth. J. Cell Biol. 118, 663-670.

Sano, K., Tanihara, H., Heimark, R.L., Obata, S., Davidson, M., St. John, T., Taketani, S., & Susuki, S. (1993). Protocadherins: A large family of cadherin-ralted molecules in central nervous system. EMBOJ 12, 2249-2256.

Shafer, S., Troyanovsky, S.M., Heid, H.W., Eshkind, L.G., Koch, P.J., & Franke, W.W. (1993). Cytoskeletal Architecture and Epithelial Differentiation: Molecular Determinants of Cell Interaction and Cytoskeletal Anchorage. C.R. Acad. Sci. Paris 316, 1316-1323.

Sheu, H.M., Kitajima, Y., & Yaoita, H. (1989). Involvement of protein kinase C in translocation of desmoplakins from cytosol to plasma membrane during desmosome formation in human squamous cell carcinoma cells grown in low to normal calcium concentration. Exp. Cell Res. 185, 176-90.

Shore, E.M., & Nelson, W.J. (1991). Biosynthesis of the cell adhesion molecule uvomorulin (E-cadherin) in Madin-Darby canine kidney epithelial cells. J. Biol. Chem. 266, 19672-80.

Skalli, O., Jones, J.C.R., Gagescu, R., & Goldman, R.D. (1994). IFAP300 is common to desmosomes and hemidesmosomes and is a possible linker of intermediate filaments to these junctions. J. Cell Biol. 125, 159-170.

Stappenbeck, T., & Green, K.J. (1992). The desmoplakin carboxyl terminus coaligns with and specifically disrupts intermediate filament networks when expressed in cultured cells. J. Cell Biol. 116, 1197-1209.

Su, L.K., Vogelstein, B., & Kinzler, K.W. (1993). Association of the APC tumor suppressor protein with catenins. Science 262, 1734-11737.

Suhrbier, A., & Garrod, D.R. (1986). An investigation of the molecular components of desmosomes in epithelial cells of five vertebrates. J. Cell Sci. 81, 223-242.

Takeichi, M. (1991). Cadherin cell adhesion receptors as a morphogenetic regulator. Science 251, 1451-1455.

Theis, D., Koch, P.J., & Franke, W.W. (1993). Differential synthesis of type 1 and type 2 desmocollin mRNAs in human stratified epithelia. Int. J. Dev. Biol. 37, 101-110.

Tonks, N.K. (1992). *Protein Tyrosine Phosphatases: The Problems of a Growing Family.* Cold Spring Harbor Symposia on Quantitative Biology, Cold Spring Harbor Laboratory Press.

Troyanovsky, S.M., Eshkind, L.G., Troyanovsky, R.B., Leube, R.E., & Franke, W.W. (1993). Contributions of cytoplasmic domains of desmosomal cadherins to desmosome assembly and intermediate filament anchorage. Cell 72, 561-574.

Tsukita, S., Oishi, K., Akiyama, T., Yamanashi, Y., & Yamamoto, T. (1991). Specific proto-oncogene tyrosine kinases of a src family are enriched in cell-to-cell adherens junctions where the level of tyrosine phosphorylation is elevated. J. Cell Biol. 113, 867-879.

Tsukita, S., & Tsukita, S. (1985). Desmocalmin: A calmodulin-binding high molecular weight protein isolated from desmosomes. J. Cell Biol. 101, 2070-2080.

Tsukita, S., & Tsukita, S. (1989). A new 82kD barbed end capping protein Radixin localized in the cell cell adherens junction-purification and characterization. J. Cell Biol. 108, 2369-2382.

Tsukita, S., & Tsukita, S. (1993). Molecular linkage between cadherins and actin filaments in cell-cell adherens junctions. Curr. Op. Cell Biol. 4, 834-839.

Volberg, T., Geiger, B., Dror, R., & Zick, Y. (1991). Modulation of intercellular adherens-type junctions and tyrosine phosphorylation of their components in RSV-transformed cultured chick lens cells. Cell Reg. 2, 105-120.

Warren, S.L., & Nelson, W.J. (1987). Nonmitogenic morphoregulatory action of pp60v-src on multicellular epithelial structures. Mol. Cell Biol. 7, 1326-37.

Wheeler, G.N., Parker, A.E., Thomas, C.L., Ataliotis, P., Poynter, D., Arnemann, J., Rutman, A.J., Pidsley, S.C., Watt, F.M., Rees, D.A., et al. (1991). Desmosomal glycoprotein DG1, a component of intercellular desmosome junctions, is related to the cadherin family of cell adhesion molecules. Proc. Natl. Acad. Sci. USA 88, 4796-800.

Wiche, G., Krepler, R., Artlieb, U., Pytela, R., & Aberer, W. (1984). Identification of plectin in different human cell types and immunolocalization at epithelial basal cell surface membranes. Exp. Cell Res. 155, 43-9.

NEURAL CELL ADHESION MOLECULES OF THE IMMUNOGLOBULIN SUPERFAMILY

John J. Hemperly

Advances in Molecular and Cell Biology,
Volume 16, pages 137-157.

ISBN: 0-7623-0143-0.

I. INTRODUCTION

The identification of the cell surface molecules found in the nervous system has recently been, and continues to be, the subject of intense scientific investigation in cell and molecular biology. The identification of such molecules is being closely followed by the determination of their structures, functions, and interactions, both during embryonic development and in the mature organism. At a somewhat slower rate, the role of these molecules in functional disorders is being elucidated. In the future, perhaps, the knowledge we gain will be applicable to the diagnosis and treatment of human disease.

Many of the known brain surface antigens were originally detected as simply that: the targets of antibodies, either polyclonal or monoclonal, raised against crude neuronal tissue preparations or isolated cell and cell lines. In other cases, the proteins were originally identified by their involvement in a biological function or process such as cell-cell adhesion (Thiery et al., 1977) or neurite outgrowth (Rathjen et al., 1987). Yet other molecules, for example those involved in the transmission of nerve impulses, were identified through the effects of a wide variety of pharmacological agents.

As these molecules of the brain began to be analyzed at the structural level using recombinant DNA techniques, it became clear that some of the molecules shared similarities with each other and with immunoglobulins. Because of this similarity, they are called members of the immunoglobulin (Ig) superfamily. It also appears to be a common feature of these molecules that they are capable of mediating cell adhesion either *in vivo* or *in vitro*. Because of these two characteristics, I will refer to this particular class of molecules in this chapter as neural Ig-CAMs.

This chapter will review some of the structural features of these Ig-CAMs, some of their more recently described interactions with other molecules of the same and different types, and some aspects of the intracellular signalling mediated through this binding. Some of these topics have been reviewed recently, both as special reviews of selected members of the Ig superfamily (Brummendorf and Rathjen, 1993; Grumet, 1992; Sonderegger and Rathjen, 1992; Yoshihara et al., 1991) or of neural cell adhesion molecules in general (Edelman and Crossin, 1991; Grumet, 1991; Rutishauser, 1993).

II. STRUCTURAL CHARACTERISTICS OF NEURAL IG-CAMS

Three molecules in particular seem to typify the neuronal Ig-CAMs: the neural cell adhesion molecule N-CAM, the L1 antigen, and contactin/F3/F11. N-CAM was the first neural Ig-CAM to be characterized in detail and is extremely well conserved among species. The L1 antigen was detected originally in mouse as the antigen recognized by a monoclonal antibody, and its rat homolog NILE was detected through its induction by nerve growth factor. Ng-CAM, another neural Ig-CAM from chick is similar to the mouse L1 antigen in many ways including domain structure and tissue localization, but shows more interspecies sequence divergence than among the other Ig-CAMs. For this reason, it has been argued that they may not be direct homologues (Mauro et al., 1992), but they will be grouped together for this discussion. The chick G4 and 8D9 antigens appear to be identical to Ng-CAM. The chicken contactin and F3 proteins seem to be identical to each other and highly homologous to the mouse F3 protein (Ranscht, 1988; Brummendorf et al., 1989; Gennarini et al., 1989). The sequence of human contactin has been reported recently (Reid et al., 1994). Other members of the Ig-CAM family include Tag-1/axonin-1 (Furley et al., 1990; Zuellig et al., 1992), Nr-CAM/Bravo (Grumet et al., 1991; de la Rosa et al., 1990), and neurofascin (Volkmer et al., 1992). Some of the molecular properties of these molecules are shown in Table 1.

Table 1. Neural Cell Adhesion Molecules of the Immunoglobulin Superfamily

Ig-CAM	*Species Homologs*	*Ig Domain*	*FnIII Domain*	*Membrane Association(s)*
N-CAM		5	2	transmembrane, GPI-linked, soluble (cDNA)
L1 antigen	NILE (Ng-CAM, G4, 8D9)	6	5	transmembrane
Nr-CAM/Bravo		6	5	transmembrane
Neurofascin		6	5	transmembrane
Ankyrin-binding glycoproteins		6	5	transmembrane
contactin/F3/F11		6	4	GPI-linked
axonin/TAG-1		6	4	transmembrane, GPI-linked

A. Immunoglobulin-like Domains

One of the distinguishing characteristics of immunoglobins and Ig-like molecules is a protein structure comprised of multiple domains. The domains themselves, about 100 amino acids in length, consist of multiple strands of beta-pleated sheet stabilized in most cases by an intradomain disulfide bond. The cysteine residues that comprise the bonds are spaced at a conserved distance of 50-70 amino acids and are flanked by a few characteristic residues (Williams, 1987). This conservation is not particular obvious and only upon alignment of the multiple domains can it be recognized. Between the cysteines, the conservation of amino acids is even less, and is not generally detectible by similarity searching programs. A detailed analysis of a number of members of the family led to the suggestion of a specific subclass, the C2-set, which is the most common form of Ig-like domain found in neural cell adhesion molecules. The C2-set Ig-CAMs resemble the classic V(ariable) domains of immunoglobulins in two of the beta strands (E and F), but they are not able to contain the loop between beta-strands (C and D) characteristic of V domains (Williams, 1987).

B. Fibronectin Type-III Repeats

Another structural characteristic shared by many neural Ig-CAMs is the fibronectin type III (FnIII) repeat. This structure, which occurs many times in the extracellular matrix protein fibronectin, appears to form a structural domain, similar in many ways to an Ig-like domain. Although the three dimensional structure of a neural Ig-CAM has not yet been determined, NMR (Baron et al., 1992) and X-ray crystallographic studies (Leahy et al., 1992) on a FnIII-like repeat of the extracellular matrix protein tenascin, have revealed a compact globular structure. Presumably, as for Ig-like domains, the residues involved in functional interactions such as cell-cell adhesion are located on the exposed surfaces of the domain with the remainder of the domain forming a structural scaffold. The corollary of this is that the Ig- and FnIII-like domains, although structurally similar, could have quite different binding activities.

C. Membrane Association

The neural Ig-CAMs discussed here are all associated with the cell surface. For L1 antigen/Ng-CAM and some of the isoforms of N-

CAM, this is achieved through hydrophobic membrane-spanning regions that divide the molecules into large, extracellular segments containing the Ig- and FnIII-domains, and intracellular segments of various sizes. Contactin/F3/F11 and one of the isoforms of N-CAM, in contrast, are associated with the membrane via a glycan-containing phosphotidylinositol-linked anchor at their carboxyl termini. Clearly these molecules can be capable of differential interactions with the inside of the cell, for example, by binding to the cytoskeleton. As will be discussed below, however, even the molecules that do not span the membrane appear able to transmit cellular signals. The mechanism for how they do this is unknown. It is also unknown why the lipid anchored molecules are relatively difficult to extract with the mild, non-ionic detergents that readily remove the transmembrane molecules. Perhaps they form complexes with other membrane components which help to anchor them securely, perhaps even to the cytoskeleton.

D. Post-translational Modification

All of the neural Ig-CAMs are glycoproteins and subject to post-translational modification. Perhaps the most unusual of these is the addition to N-CAM of a developmentally regulated alpha-2,8-polysialic acid. This modification, which in higher vertebrates seems to be largely, but not totally, specific to N-CAM, has a profound effect on the binding ability of N-CAM. For example, the highly polysialylated form of N-CAM, which is found more often in embryonic nervous tissue and where the polysialic acid can comprise up to 30% of the weight of the molecule, binds less avidly than the less glycosylated, adult form. It is also interesting in that the few places in the adult where the polysialic acid-rich form of N-CAM persists, such as the olfactory bulb, neuronal remodelling continues after birth. It has also been suggested that the presence of the polysialic acid can modulate the distance between cells and influence the interactions of other cell surface, non-CAM components (Rutishauser et al., 1988).

Neural Ig-CAMs can also be phosphorylated, almost always on serine or threonine. Although the significance of the phosphorylation is as yet unknown, it has been demonstrated that two particular protein kinases can phosphorylate N-CAM (Mackie et al., 1989) and that a protein kinase can co-purify with L1 antigen (Sadoul et al., 1989).

E. Generation of Protein Isoforms

The synthesis and processing of Ig-CAMs can be very complex. For example, N-CAM is particularly rich in alternative forms that arise by alternate splicing of RNA transcribed from a single genetic locus (Cunningham et al., 1987). Through this alternative splicing can arise molecular isoforms that differ in their attachment (or lack of attachment to the cell surface), in their presentation on the cell surface, and on their post-translation modifications both extracellular and intracellular. It has also been demonstrated that very small differences that arise by alternate splicing in the fourth Ig-like domain (the VASE exon), can have striking effects on the ability of N-CAM to mediate neurite outgrowth (Doherty et al., 1992). The factors that regulate this splicing as well as the expression and function of neural Ig-CAMs are being investigated (Walsh and Doherty, 1993), but are still largely undefined. For example, there has been a detailed analysis of the nucleotide sequences involved in the splicing of the large intracellular intron of N-CAM (Tacke and Goridis, 1991), but there have as yet been no reports of the cellular factors influencing the splicing patterns. The factors controlling the expression of neural Ig-CAMs are also under investigation. For example, the effects of a number of cytokines and growth factors on N-CAM and L1 expression have been assessed (Perides et al., 1994). As mentioned above, the L1 antigen in the rat was originally isolated as NILE, the **N**erve growth factor-**I**nducible **L**arge **E**xternal antigen of PC-12 pheochromocytoma cells (McGuire et al., 1978). A particular interesting recent observation is that the N-CAM gene is a target of several homeobox proteins, consistent with its highly regulated expression during development (Hirsch et al., 1991; Jones et al., 1992).

III. LOCALIZATION OF THE NEURAL IG-CAMS

The members of the neuronal Ig superfamily can be distinguished and classified in a number of ways. Sometimes it is useful to distinguish them in terms of their prevalence and order of appearance during embryonic development. For example, N-CAM is found very early in embryonic development, perhaps as early as the two-cell stage and certainly as early as gastrulation. It seems to be involved in establishing the body plan of the embryo. In contrast, the molecules L1/Ng-CAM and contactin/F3/F11 arise later in development and may not reach significant levels until after birth.

Sometimes it is useful to distinguish between Ig-superfamily molecules that are relatively specific to the nervous system and those with a wider tissue distribution. For example, N-CAM, although originally detected and widely distributed in the nervous system, is also found on a number of neuroectodermal and neuroendodermal derivatives such as cells of the adrenal glands, skeletal, cardiac, and smooth muscle. It can occasionally be expressed by endothelial cells (Brandon et al., 1993). It is also found on NK cells and certain T-lymphocytes, where it has been termed the CD56 antigen and is used as a marker for these cell types (Lanier et al., 1989). N-CAM is also a major antigen of smooth cell lung carcinoma; in antibody classification workshops, it is recognized as the "cluster 1" antigen (Patel et al., 1989). In contrast, L1 antigen/Ng-CAM appear to be somewhat more specific, although there have been recent reports of L1 antigen in adrenal medulla (Poltorak and Freed, 1990), and on many lymphocytes in mice (Kowitz et al., 1992), and in lymphoid tumors (Kowitz et al., 1993). However, L1 does not seem to be expressed on human peripheral blood cells (unpublished results). So far, contactin/F3/F11 is very neuronally specific, although small amounts have been found in lung and pancreas (Reid et al., 1994).

It is also possible to distinguish the neural Ig superfamily by their subcellular distributions (Brummendorf and Rathjen, 1993). Some Ig-CAMs, such as the L1 antigen are localized more to axon than cell bodies and dendrites. This reveals itself as a prominent staining of fiber tracts in the brain. Other Ig-CAMs, such as N-CAM, are located on cell soma, dendrites, and axons, although they appear to be particularly concentrated at synapses (Persohn et al., 1989). All appear to be concentrated at areas of cell-cell contact with other Ig-CAM-bearing cells. For example, in peripheral nerve, N-CAM and L1 antigen are found between Schwann cells and axons, but much less so on the Schwann cell surfaces exposed to extracellular matrix (Martini, 1994). It has also been suggested that phosphatidylinositol-linked molecules may be preferentially targeted to axons, but in fact the distribution of such molecules may be highly complex, depending on the neural cell type, the Ig-CAM molecule itself, and the differentiation state of the cell (Faivre-Sarrailh and Rougon, 1993).

IV. IG-CAM BINDING AND LIGANDS

As mentioned previously, one of the first questions asked of neural Ig-CAMs is "Are they capable of mediating cellular adhesion?" In all cases, the answer has turned out to be "yes," although sometimes rather extreme assays are required to demonstrate the binding. In contrast to some of the other adhesion molecules, such as the calcium-dependent cadherins, the Ig-CAMs generally show much weaker binding. It is often the case, in fact, that the experimentalist may wonder if the primary function of these molecules is adhesion at all (Zanetta, 1991), or if the adhesion might reflect instead a cell signalling function of which adhesion is a necessary but not sufficient part (see following). Be that as it may, many efforts have been made to study Ig-CAM binding. In particular, these studies have generally focused on the interaction of neural Ig-CAMs with (1) cytoskeletal elements within the cell, or (2) with extracellular ligands on the same ("cis") or other ("trans") cell surfaces.

The data on the binding of Ig-CAMs to intracellular elements is rather sparse, but because such an interaction would provide a mechanism for communicating signals from outside to inside the cell or vice versa, it remains an attractive hypothesis. Early reports demonstrated that some forms of N-CAM are restricted in their lateral mobility on the cell surface (Pollerberg et al., 1986) presumably through their interaction with cytoskeletal elements. There was also a demonstration of binding of N-CAM to spectrin *in vitro* (Pollerberg et al., 1987), but there has not been extension of these interesting observations since then. Recently, however, it has been demonstrated that ankyrin, a well-characterized component of the cytoskeleton can bind a molecule related to neurofacscin, another member of Ig-CAM family (Davis et al., 1993). These studies began at the other end, inside the cell, asking what molecules interact with ankyrin. A clear and reasonable stoichiometry of binding was demonstrated. Also, because the cytoplasmic amino acid sequences of Ig-CAMs are often the most highly conserved regions of the molecule, it is tempting to speculate that other Ig-CAMs will show similar interactions with ankryn or other cytoskeletal proteins.

Early studies with N-CAM revealed that purified N-CAM is capable of binding to N-CAM molecules on other cells or lipid vesicles in what was called a "homophilic" interaction, that is, N-CAM binding to N-CAM. In contrast, another Ig-CAM, Ng-CAM,

Table 2. Interactions of the Neural Ig-CAMs

Neural Ig-CAM	*Ligand*
N-CAM	Homophilic
	N-CAM 3rd Ig domain (Rao et al., 1993)
	Heterophilic:
	proteoglycan (Cole and Burg, 1989)
	spectrin (Pollerberg et al., 1987)
	collagen (Probstmeier et al., 1992)
	L1 antigen (Kadmon et al., 1990)
	unknown (Murray and Jensen, 1992)
L1/Ng-CAM	Homophilic:
	L1/Ng-CAM (Lemmon et al., 1989; Miura et al., 1992)
	Heterophilic:
	unknown (Grumet and Edelman, 1988)
	axonin-1 (Kuhn et al., 1991)
	contactin/F3/F11
	extracellular matrix (Werz and Schachner, 1988)
contactin/F3/F11	Homophilic:
	(weak) (Gennarini et al., 1991)
	Heterophilic:
	Nr-CAM/Bravo (Morales et al., 1993)
	Restrictin (Rathjen et al., 1991)
	J1-160/180=janusin (Pesheva et al., 1993)
	Ng-CAM (Brummendorf et al., 1993)
	tenascin/cytotactin (Zisch et al., 1992)
Nr-CAM	Homophilic:
	Nr-CAM (Mauro et al., 1992)
	Heterophilic:
	Unknown (Mauro et al., 1992)
axonin-1/TAG-1	Homophilic:
	axonin-1 (Rader et al., 1993)

was clearly involved in neuron-glia binding but was localized on neurons and absent from the glia used in these assays. This implied there must be some glial, non-Ng-CAM ligand, and at least in some cases, Ng-CAM binding must be "heterophilic." More recently, it has become clear that many neural Ig-superfamily members are capable of multiple interactions, some homophilic and some heterophilic (Table 2). In addition, these molecules can bind not only to other molecules on the same or other cell surfaces, but also to molecules of the extracellular matrix. In fact, some of the neural Ig-CAMs appear to be released from the cell surface and localize to the extracellular matrix itself (Rieger et al., 1988).

Probably one of the more intriguing but difficult to address issues of CAM-CAM binding involves the interactions between CAMs on the same plasma membrane. Such interactions are clearly indicated by the synergistic effects of multiple CAMs on cell-cell binding (Kadmon, 1990) and by chemical cross-linking experiments (Simon et al., 1991). More recently, it has been demonstrated that some "cis" interactions may be mediated by carbohydrates, such that lectin-like regions on one CAM bind to carbohydrates on other CAMs (Horstkorte et al., 1993). Such "cis" interactions would also be consistent with ability in some cases, for antibodies to one CAM to block cell-cell adhesion although multiple CAMs are present.

The interaction between CAMs on different membranes is somewhat easier to demonstrate. As mentioned above, N-CAM to N-CAM binding in reconstituted lipid vesicles provided evidence not only for homophilic binding but for the extreme dependence of the binding on CAM concentration (Hoffman and Edelman, 1983). Since then, many of the Ig-CAMs have been transfected into cell lines to generate cell-cell adhesion phenomena, and in most cases, to generate the ability to support neurite outgrowth. In fact, this has become one of the primary early requirements in the characterization of a new Ig-CAM. By such methods, even Thy-1, an Ig-CAM comprising a single immunoglobulin domain, can mediate adhesion (Doherty et al., 1993).

Because of recombinant DNA approaches, it has been possible to begin to map some Ig-CAM binding domains. For example, for N-CAM, there appears to be a region in the third immunoglobulin-like domain that is involved in homophilic binding, and synthetic peptides based on this sequence can inhibit adhesion (Rao et al, 1993). In contrast, it appears to be a very highly charged sequence of amino acids in the second domain that mediates binding to an N-CAM-associated proteoglycan (Cole and Akeson, 1989). Similar studies have been performed for contactin/F3/F11, where it has been demonstrated that binding to the extracellular matrix molecule tenascin is mediated by sequences in the first two immunoglobulin-like domains (Zisch et al., 1992). Clearly, a major area of future research interest will involve the detailed mapping of the domains in terms of their ligands, and determining how the various CAM-ligand binding activities contribute to the biology of the CAMs.

V. IG-CAMS IN SIGNAL TRANSDUCTION

As mentioned earlier, all Ig-CAMs can mediate cell-cell adhesion in some circumstances, such as among transfected cell lines. However, attempts to measure the binding affinity constants for CAMs in solution have been difficult. Using molecules released from the cell surface with detergents, binding constants for homophilic binding of adult and embryonic N-CAM were measured to be 6.9×10^{-8} M and 1.23×10^{-6} M, respectively (Moran and Bock, 1988). This weakness of binding, combined with early observations that antibodies to the Ig-CAM known as Thy-1, could deliver a mitogenic signal, suggests that the primary function of these molecules may in fact not be binding, but the transduction of signals from the outside to the inside of the cell. More recent studies have looked at the effect of antibodies to CAMs and of CAM fragments on signaling pathways and cellular responses.

Attempts to study signal transduction via Ig-CAMs have generally followed two main paths. First, the effects of endogenously applied CAMs on biological function such as neurite extension or cell migration have been measured and the response of such functions to pharmacological agents has been used to deduce molecular mechanisms. Alternatively, the direct effects of CAM binding on biochemical events such as protein phosphorylation or ion and second messenger levels have been measured.

Early studies focused on the ability of CAM substrates to facilitate or impede neurite outgrowth. From these studies came the general conclusion that adhesivity itself did not correlate directly with neurite outgrowth. There seemed to be a balance between too tight binding that led to poor outgrowth (a flypaper effect) or insufficient adhesivity (the greased pole effect) that was similarly ineffectual at supporting neurite outgrowth. It was only at intermediate binding levels that neurites were generated efficiently. However, this did not really address the issue of what changes were taking place in the growing cells. Also, some groups argued that CAMs attached to dishes, while clearly having differential effects, could not be very useful models of what was happening at the cell surface. For example, such systems could not easily allow for "cis" interactions of the CAMs themselves nor could there be communication via the cytoskeleton. Arguably, the foremost proponents of this view are Walsh, Doherty, and their colleagues, who have established transfected cell systems for studying neurite outgrowth. Originally these systems studied neurite outgrowth from cell lines such as the PC12 rat pheochromocytoma or primary neurons on fibroblasts transfected with N-CAM, N-cadherin, or L1 antigen. It was clear that the transfected CAMs could support neurite outgrowth, although the cell adhesion in this paradigm was largely independent of the transfected CAM. This again suggests that binding *per se* may be mediated by other molecules, most likely integrins and their ligands, whereas signalling resulting, in this case, in neurite outgrowth may be mediated by neural Ig-CAMs. These workers have then been able to perturb outgrowth with various pharmacological agents to demonstrate in their paradigm that neurite outgrowth is dependent on L-type and N-type calcium channels and pertussis toxin sensitive G proteins (Doherty et al., 1991). Moreover, the neurite outgrowth can be mimicked by stimulation of these systems alone (Saffell et al., 1992).

Given these phenomena, the effects of CAMs and CAM isoforms on neurite outgrowth were compared to show that at least N-CAM, N-cadherin, and L1 antigen mediated neurite outgrowth rely on similar mechanisms. Also, it was possible to demonstrate that the VASE-containing isoform of N-CAM is less able to support outgrowth as is the isoform with large intracellular, presumptively cytoskeletal-binding domain (Walsh et al., 1992). In both cases there appears to be a correlation with the biology in that both the VASE isoform and 180kD isoform increase in prevalence during development as neurite outgrowth decrease *in vivo*.

Alternatively, the effects of CAM binding on intracellular substrates have been studied directly. For examples, antibodies and fragments of N-CAM and L1 can cause changes in intracellular calcium levels (Schuch et al., 1989; Williams et al., 1992; Asou et al., 1992; von Bohlen und Halbach et al., 1992). They can also lead to changes in phosphorylation patterns in isolated nerve growth cones, apparently due to differential effects on cellular kinases and phosphatases (Atashi et al., 1992).

Clearly there are multiple, not necessarily exclusive, mechanisms for such signal transduction. As mentioned above, there could be direct signalling through the cytoskeleton. Alternatively, there could be signal transduction through Ig-CAMs either alone, or more likely, through interaction complexes (Rutishauser, 1993; Doherty and Walsh, 1992).

VI. IG-CAMS IN DISEASE

Are Ig-CAMs involved in disease? There appear to be two major areas in which disturbances of these molecules could result in disease: during development and postnatally, where differences in adhesion or cell signalling could lead to aberrant cell behavior. In fact, a number of recent reports suggest that neural Ig-CAMs may underlie developmental disturbances. For example, transgenic mice have been generated that are null mutants for either a neuronally restricted isoform of N-CAM (Tomasiewicz et al., 1993) or for all isoforms of N-CAM (Cremer et al., 1994). These transgenic animals demonstrate that animals with disrupted IgCAMs can still be viable and fertile. However, the animals are clearly **not** normal and show structural abnormalities of the brain and impaired performance in learning tests. The recent identification of L1 antigen as the disturbed molecule in the human disease X-linked hydrocephaly (Rosenthal et al., 1992; Van Camp et al., 1993; Jouet et al., 1993), similarly points to defects in a particular Ig-CAM leading to developmental disturbances. The human gene encoding the human homolog of Tag-1, known as Tax-1 (Tsiotra et al., 1993), has been linked to chromosome 1, in a region implicated in microcephaly and Van der Woude syndrome (Kenwrick et al., 1993), and the gene encoding an additional N-CAM-like molecule has been implicated in Kallmann syndrome (Legouis et al., 1991; Franco et al., 1991). In the latter syndrome, there appears to be a failure of neurons to migrate properly during development and it is tempting to speculate that Ig-CAMs are involved in this migration. Another Ig-CAM, although not discussed in detail in this chapter, is the P_0 protein of peripheral myelin. Defects in P_0 have been implicated in some forms of Charcot-Marie-Tooth neuropathies (Hayasaka et al., 1993). Future studies both in human genetic disease and in experiment animal models of targeted gene disruption (Schachner, 1993) should provide additional insight into Ig-CAM mechanisms in disease.

The finding of Ig-CAMs as tumor antigens suggest they may be indicative or causative of altered cellular behavior occurring post-embryonically. It has been recognized for some time that Ig-CAMs can be found immunohistologically in a number of human tumors (Garin-Chesa et al., 1991; Jin et al., 1991; Molenaar et al., 1991). It has recently been suggested that serum N-CAM levels may be indicative of multiple myeloma (Kaiser et al., 1994) or small cell lung carcinoma (Jaques et al., 1993). There are still other reports that Ig-CAMs can determine metastatic preferences—lymphoma expressing N-CAM has an increased tendency to move to the brain (Kern et al., 1992). Alternatively, it has been suggested that N-CAM positive multiple myeloma is correlated with a better prognosis (Van Camp et al., 1990). Although this has not been demonstrated as clearly with neural Ig-CAMs, failures in the function of Ig-superfamily CAMs in the immune system is correlated with disfunction and is being investigated as a point for therapeutic intervention.

What appears to be emerging is a family of molecules, the neural Ig-CAMs, with related molecular structures that are capable of a much wider range of functions and ligands than initially appreciated. The Ig-CAMs not only mediate adhesion, but also signal transduction. In fact, it could well be that the signalling is of much more profound consequence than the adhesion, which could be mediated by much more robust adhesion proteins such as the cadherins or integrins. It is also becoming clearer that Ig-CAMs have a multitude of ligands, perhaps with specific binding mediated by distinct structural domains. For example, N-CAM may bind N-CAM using the third Ig domain, yet bind to proteoglycans or heparin via amino acids in the second Ig domain. It will certainly take even more efforts to sort out these various binding and signalling interactions and determine the detailed functions of these molecules *in vivo*.

REFERENCES

Asou, H., Miura, M., Kobayashi, M., Uyemura, K., & Itoh, K. (1992). Cell adhesion molecule L1 guides cell migration in primary reaggregation cultures of mouse cerebellar cells. Neurosci. Lett. 144, 221-224.

Atashi, J. R., Klinz, S. G., Ingraham, C. A., Matten, W. T., Schachner, M., & Maness, P. F. (1992). Neural cell adhesion molecules modulate tyrosine phosphorylation of tubulin in nerve growth cone membranes. Neuron 8, 831-842.

Baron, M., Main, A. L., Driscoll, P. C., Mardon, H. J., Boyd, J., & Campbell, I. D. (1992). H-1 NMR assignment and secondary structure of the cell adhesion type III module of fibronectin. Biochemistry 31, 2068-2073.

Brandon, E. P., Kotik, A. N., Vachino, G., Swack, J. A., Libby, P., Janicka, M. W., Atkins, M. B., & Mier, J. W. (1993). Cultured human endothelial cells express the 140-kD isoform of Neural Cell Adhesion Molecule (NCAM). Internat. J. Immunopath. Pharmacol. 6, 77-84.

Brummendorf, T., & Rathjen, F. G. (1993). Axonal glycoproteins with immunoglobulin and fibronectin type-III-related domains in vertebrates—Structural features, binding activities, and signal transduction. J. Neurochem. 61, 1207-1219.

Brummendorf, T., Hubert, M., Treubert, U., Leuschner, R., Tarnok, A., & Rathjen, F. G. (1993). The axonal recognition molecule F11 is a multifunctional protein—Specific domains mediate interactions with Ng-CAM and restrictin. Neuron 10, 711-727.

Brummendorf, T., Wolff, J. M., Frank, R., & Rathjen, F. G. (1989). Neural cell recognition molecule F11: Homology with fibronectin type III and immunoglobulin type C domains. Neuron 2, 1351-1361.

Cole, G. J., & Akeson, R. (1989). Identification of a heparin binding domain of the neural cell adhesion molecule N-CAM using synthetic peptides. Neuron 2, 1157-1165.

Cole, G. J., & Burg, M. (1989). Characterization of a heparan sulfate proteoglycan that copurifies with the neural cell adhesion molecule. Exp. Cell Res. 182, 44-60.

Cremer, H., Lange, R., Christoph, A., Plomann, M., Vopper, G., Roes, J., Brown, R., Baldwin, S., Kraemer, P., Scheff, S., Barthels, D., Rajewsky, K., & Wille, W. (1994). Inactivation of the N-CAM gene in mice results in size reduction of the olfactory bulb and deficits in spatial learning. Nature (London) 367, 455-459.

Cunningham, B. A., Hemperly, J. J., Murray, B. A., Prediger, E. A., Brackenbury, R., & Edelman, G. M. (1987). Neural cell adhesion molecule: Structure, immunoglobulin-like domains, cell surface modulation, and alternative RNA splicing. Science (Washington) 236, 799-806.

Davis, J. Q., Mclaughlin, T., & Bennett, V. (1993). Ankyrin-binding proteins related to nervous system cell adhesion molecules—Candidates to provide transmembrane and intercellular connections in adult brain. J. Cell Biol. 121, 121-133.

de la Rosa, E. J., Kayyem, J. F., Roman, J. M., Stierhof, Y-D, Dreyer, W. J., & Schwarz, U. (1990). Topologically restricted appearance in the developing chick retinotectal system of Bravo, a neural cell surface protein: Experimental modulation by environmental cues. J. Cell Biol. 111, 3087-3096.

Doherty, P., & Walsh, F. S. (1992). Cell adhesion molecules, second messengers, and axonal growth. Curr. Opin. Neurobiol. 2, 595-601.

Doherty, P., Ashton, S. V., Moore, S. E., & Walsh, F. S. (1991). Morphoregulatory activities of NCAM and N-cadherin can be accounted for by G protein-dependent activation of L- and N-type neuronal Ca^{2+} channels. Cell 67, 21-33.

Doherty, P., Moolenaar, C. E. C. K., Ashton, S. V., Michalides, R. J. A. M., & Walsh, F. S. (1992). The VASE exon downregulates the neurite growth-promoting activity of NCAM 140. Nature (London) 356, 791-793.

Doherty, P., Singh, A., Rimon, G., Bolsover, S. R., & Walsh, F. S. (1993). Thy-1 antibody-triggered neurite outgrowth requires an influx of calcium into neurons via N-type and L-type calcium channels. J. Cell Biol. 122, 181-189.

Durbec, P., Gennarini, G., Goridis, C., & Rougon, G. (1992). A soluble form of the F3 neuronal cell adhesion molecule promotes neurite outgrowth. J. Cell Biol. 117, 877-887.

Edelman, G. M., & Crossin, K. L. (1991). Cell adhesion molecules—Implications for a molecular histology. Annu. Rev. Biochem. 60, 155-190.

Faivre-Sarrailh, C., & Rougon, G. (1993). Are the glypiated adhesion molecules preferentially targeted to the axonal compartment? Mol. Neurobiol. 7, 49-60.

Franco, B., Guioli, S., Pragliola, A., Incerti, B., Bardoni, B., Tonlorenzi, R., Carrozzo, R., Maestrini, E., Pieretti, M., Taillon-Miller, P., Brown, C.J., Willard, H.F., Lawrence, C., Persico, M.G., Camerino, G., & Ballabio, A. (1991). A gene deleted in Kallmann's syndrome shares homology with neural cell adhesion and axonal path-finding molecules. Nature (London) 353, 529-536.

Furley, A. J., Morton, S. B., Manalo, D., Karagogeos, D., Dodd, J., & Jessell, T. M. (1990). The axonal glycoprotein TAG-1 is an immunoglobulin superfamily member with neurite outgrowth-promoting activity. Cell 61, 157-170.

Garin-Chesa, P., Fellinger, E. J., Huvos, A. G., Beresford, H. R., Melamed, M. R., Triche, T. J., & Rettig, W. J. (1991). Immunohistochemical analysis of neural cell adhesion molecules—differential expression in small round cell tumors of childhood and adolescence. Am. J. Pathol. 139, 275-286.

Gennarini, G., Cibelli, G., Rougon, G., Mattei, M.-G., & Goridis, C. (1989). The mouse neuronal cell surface protein F3: A phophatidylinositol-anchored member of the immunoglobulin superfamily related to chicken contactin. J. Cell Biol. 109, 775-788.

Gennarini, G., Durbec, P., Boned, A., Rougon, G., & Goridis, C. (1991). Transfected F3/F11 neuronal cell surface protein mediates intercellular adhesion and promotes neurite outgrowth. Neuron 6, 595-606.

Grumet, M. (1991). Cell adhesion molecules and their subgroups in the nervous system. Curr. Opin. Neurobiol. 1, 370-376.

Grumet, M. (1992). Structure, expression, and function of Ng-CAM, a member of the immunoglobulin superfamily involved in neuron-neuron and neuron-glia adhesion. J. Neurosci. Res. 31, 1-13.

Grumet, M., & Edelman G.M. (1988). Neuron-glia cell adhesion molecule interacts with neurons and astroglia via different binding mechanisms. J. Cell Biol. 106, 487-503.

Grumet, M., Mauro, V., Burgoon, M. P., Edelman, G. M., & Cunningham, B. A. (1991). Structure of a new nervous system glycoprotein Nr-CAM and its relationship to subgroups of neural cell adhesion molecules. J. Cell Biol. 113, 1399-1412.

Hayasaka, K., Ohnishi, A., Takada, G., Fukushima, Y., & Murai, Y. (1993). Mutation of the myelin Po gene in Charcot-Marie-tooth neuropathy type 1. Biochem. Biophys. Res. Commun. 194, 1317-1322.

Hirsch, M.-R., Valarche, I., Deagostini-Bazin, H., Pernelle, C., Joliot, A., & Goridis, C. (1991). An upstream regulatory element of the NCAM promoter contains a binding site for homeodomains. FEBS Lett. 287, 197-202.

Hoffman, S., & Edelman, G. M. (1983). Kinetics of homophilic binding by embryonic and adult forms of the neural cell adhesion molecule. Proc. Natl. Acad. Sci. USA 80, 5762-5766.

Horstkorte, R., Schachner, M., Magyar, J. P., Vorherr, T., & Schmitz, B. (1993). The fourth immunoglobulin-like domain of NCAM contains a carbohydrate recognition domain for oligomannosidic glycans implicated in association with L1 and neurite outgrowth. J. Cell Biol. 121, 1409-1421.

Jaques, G., Auerbach, B., Pritsch, M., Wolf, M., Madry, N., & Havemann, K. (1993). Evaluation of serum neural cell adhesion molecule as a new tumor marker in small cell lung carcancer. Cancer 72, 418-425.

Jin, L., Hemperly, J.J., & Lloyd, R.V. (1991). Expression of neural cell adhesion molecule in normal and neoplastic human neuroendocrine tissues. Am. J. Pathol. 138, 961-969.

Jones, F. S., Prediger, E. A., Bittner, D. A., De Robertis, E. M., & Edelman, G. M. (1992). Cell adhesion molecules as targets for *Hox* genes: Neural cell adhesion molecule promoter activity is modulated by cotransfection with *Hox-2.5* and *-2.4*. Proc. Natl. Acad. Sci. USA 89, 2086-2090.

Jouet, M., Rosenthal, A., Macfarlane, J., Kenwrick, S., & Donnai, D. (1993). A missense mutation confirms the L1 defect in X-linked hydrocephalus (HSAS). Nature Genetics. 4, 331.

Kadmon, G., Kowitz, A., Altevogt, P., & Schachner, M. (1990). The neural cell adhesion molecule N-CAM enhances L1-dependent cell-cell interactions. J. Cell Biol. 110, 193-208.

Kaiser, U., Jaues, G., Havemann, K., & Auerbach, B. (1994). Serum N-CAM: A potential new prognostic marker for multiple myeloma. Blood 83, 871-873.

Kenwrick, S., Leversha, M., Rooke, L., Hasler, T., & Sonderegger, P. (1993). Localization of the human TAX-1 gene to 1q32.1—A region implicated in microcephaly and Van der Woude syndrome. Human Molecular Genetics. 2, 1461-1462.

Kern, W. F., Speir, C. M., Hanneman, E. H., Miller, T. P., Matzner, M., & Grogan, T. M. (1992). Neural cell adhesion molecule-positive peripheral T-cell lymphoma-a rare variant with a propensity for unusual sites of involvement. Blood 79, 2432-2437.

Kowitz, A., Kadmon, G., Eckert, M., Schirrmacher, V., Schachner, M., & Altevogt, P. (1992). Expression and function of the neural cell adhesion molecule L1 in mouse leukocytes. Eur. J. Immunol. 22, 1199-1205.

Kowitz, A., Kadmon, G., Verschueren, H., Remels, L., Debaetselier, P., Hubbe, M., Schachner, M., Schirrmacher, V., & Altevogt, P. (1993). Expression of L1 cell adhesion molecule is associated with lymphoma growth and metastasis. Clin. Exper. Metastasis 11, 419-429.

Kuhn, T.B., Stoeckli, E.T., Condrau, M.A., Rathjen, F.G., & Sonderegger, P. (1991). Neurite outgrowth on immobilized axonin-1 is mediated by a heterophilic interaction with L1(G4). J Cell Biol. 115, 1113-1126.

Lanier, L. L., Testi, R., Bindl, J., & Phillips, J. H. (1989). Identity of Leu-19 (CD56) leukocyte differentiation antigen and neural cell adhesion molecule. J. Exp. Med. 169, 2233-2238.

Leahy, D. J., Hendrickson, W. A., Aukhil, I., & Erickson, H. P. (1992). Structure of a fibronectin type-III domain from tenascin phased by MAD analysis of the selenomethionyl protein. Science (Washington) 258, 987-991.

Legouis, R., Hardelin, JP, Levilliers, J., Claverie, JM, Compain, S., Wunderle, V., Millasseau, P., Le, Paslier D., Cohen, D., Caterina, D., Bouguerleret, L., Delemarre-Van de Waal, H., Lutfalla, G., Weissenbach, J., & Petit, C. (1991). The candidate gene for the X-linked Kallmann syndrome encodes a protein related to adhesion molecules. Cell 67, 423-435.

Lemmon, V., Farr, K. L., & Lagenaur, C. (1989). L1-mediated axon outgrowth occurs via a homophilic binding mechanism. Neuron 2, 1597-1603.

Mackie, K., Sorkin, B. C., Nairn, A. C., Greengard, P., Edelman, G. M., & Cunningham, B. A. (1989). Identification of two protein kinases that phosphorylate the neural cell adhesion molecule, N-CAM. J. Neurosci. 9, 1883-1896.

Martini, R. (1994). Expression and functional roles of neural cell surface molecules and extracellular matrix components during development and regeneration of peripheral nerves. J. Neurocytol. 23, 1-28.

Mauro, V. P., Krushel, L. A., Cunningham, B. A., & Edelman, G. M. (1992). Homophilic and heterophilic binding activities of Nr-CAM, a nervous system cell adhesion molecule. J. Cell Biol. 119, 191-202.

McGuire, J. C., Greene L.A., & Fuvano A.V. (1978). NGF stimulates incorporation of fucose or glucosamine into an external glycoprotein in cultured rat PC12 pheochromocytoma cells. Cell 15, 357-365.

Miura, M., Asou, H., Kobayashi, M., & Uyemura, K. (1992). Functional expression of a full-length cDNA coding for rat neural cell adhesion molecule-L1 mediates homophilic intercellular adhesion and migration of cerebellar neurons. J. Biol. Chem. 267, 10752-10758.

Molenaar, W.M., de Leij L., & Trojanowski, J.Q. (1991). Neuroectodermal tumors of the peripheral and the central nervous system share neuroendocrine N-CAM-related antigens with small cell lung carcinomas. Acta Neuropathol. (Berl.) 83, 46-54.

Morales, G., Hubert, M., Brummendorf, T., Treubert, U., Tarnok, A., Schwarz, U., & Rathjen, F. G. (1993). Induction of axonal growth by heterophilic interactions between the cell surface recognition protein-F11 and protein-Nr-CAM/Bravo. Neuron 11, 1113-1122.

Moran, N., & Bock, E. (1988). Characterization of the kinetics of neural cell adhesion molecule homophilic binding. FEBS Lett. 242, 121-124.

Murray, B. A., & Jensen, J. J. (1992). Evidence for heterophilic adhesion of embryonic retinal cells and neuroblastoma cells to substratum-adsorbed NCAM. J. Cell Biol. 117, 1311-1320.

Patel, K., Moore, S. E., Dickson, G., Rossell, R. J., Beverley, P. C., Kemshead, J. T., & Walsh, F. S. (1989). Neural cell adhesion molecule (NCAM) is the antigen recognized by monoclonal antibodies of similar specificity in small-cell lung carcinoma and neuroblastoma. Int. J. Cancer 44, 573-578.

Perides, G., Safran, R. M., Downing, L. A., & Charness, M. E. (1994). Regulation of neural cell adhesion molecule and L1 by the transforming growth factor-beta superfamily—Selective effects of the bone morphogenetic proteins. J. Biol. Chem. 269, 765-770.

Persohn, E., Pollerberg, G. E., & Schachner, M. (1989). Immunoelectron-microscopic localization of the 180 kD component of the neural cell adhesion molecule N-CAM in postsynaptic membranes. J. Comp. Neurol. 288, 92-100.

Pesheva, P., Gennarini, G., Goridis, C., & Schachner, M. (1993). The F3/11 cell adhesion molecule mediates the repulsion of neurons by the extracellular matrix glycoprotein J1-160/180. Neuron 10, 69-82.

Pollerberg, G. E., Schachner, M., & Davoust, J. (1986). Differentiation state-dependent surface mobilities of two forms of the neural cell adhesion molecule. Nature (London) 324, 462-465.

Pollerberg, G. E., Burridge, K., Krebs, K. E., Goodman, S. R., & Schachner, M. (1987). The 180-kD component of the neural cell adhesion molecule N-CAM is involved in cell-cell contacts and cytoskeleton-membrane interactions. Cell Tissue Res. 250, 227-236.

Poltorak, M., & Freed, W. J. (1990). Cell adhesion molecules in adrenal medulla grafts enhancement of chromaffin cell L1-Ng-CAM expression and reorganization of extracellular matrix following transplantation. Exp. Neurol. 110, 73-85.

Probstmeier, R., Fahrig, T., Spiess, E., & Schachner, M. (1992). Interactions of the neural cell adhesion molecule and the myelin-associated glycoprotein with collagen type-I—Involvement in fibrillogenesis. J. Cell Biol. 116, 1063-1070.

Rader, C., Stoeckli, E. T., Ziegler, U., Osterwalder, T., Kunz, B., & Sonderegger, P. (1993). Cell-cell adhesion by homophilic interaction of the neuronal recognition molecule axonin-1. Eur. J. Biochem. 215, 133-141.

Ranscht, B. (1988). Sequence of contactin, a 130 kD glycoprotein concentrated in areas of interneuronal contact, defines a new member of the immunoglobulin supergene family in the nervous system. J. Cell Biol. 107, 1561-1573.

Rao, Y., Wu, X. F., Yip, P., Gariepy, J., & Siu, C. H. (1993). Structural characterization of a homophilic binding site in the neural cell adhesion molecule. J. Biol. Chem. 268, 20630-20638.

Rathjen, F. G., Wolff, J. M., Frank, R., Bonhoeffer, R., & Rutishauser, U. (1987). Membrane glycoproteins involved in neurite fasciculation. J. Cell Biol. 104, 343-353.

Rathjen, F. G., Wolff, J. M., & Chiquet-Ehrismann, R. (1991). Restrictin—a chick neural extracellular matrix protein involved in cell attachment co-purifies with the cell recognition molecule-F11. Development 113, 151-164.

Reid, R. A., Bronson, D. D., Young, K. M., & Hemperly, J. J. (1994). Identification and characterization of the human cell adhesion molecule contactin. Mol. Brain Res. 21, 1-8.

Reiger, F., Nicolet, M., Pincon-Raymond, M., Murawsky, M., Levi, G., & Edelman, G. M. (1988). Distribution and role in regeneration of N-CAM in the basal laminae of muscle and Schwann cells. J. Cell Biol. 107, 707-719.

Rosenthal, A., Jouet, M., & Kenwrick, S. (1992). Aberrant splicing of neural cell adhesion molecule L1 mRNA in a family with X-linked hydrocephalus. Nat Genet. 2, 107-112.

Rutishauser, U. (1993). Adhesion Molecules of the nervous system. Curr. Opin. Neurobiol. 3, 709-715.

Rutishauser, U., Acheson, A., Hall, A. K., Mann, D. M., & Sunshine, J. (1988). The neural cell adhesion molecule (N-CAM) as a regulator of cell-cell interactions. Science (Washington) 240, 53-57.

Sadoul, R., Kirchhoff, F., & Schachner, M. (1989). A protein kinase activity is associated with and specifically phosphorylates the neural cell adhesion molecule L1. J. Neurochem. 53, 1471-1478.

Saffell, J. L., Walsh, F. S., & Doherty, P. (1992). Direct activation of second messenger pathways mimics cell adhesion molecule-dependent neurite outgrowth. J. Cell Biol. 118, 663-670.

Schachner, M. (1993). The analysis of neural recognition molecules benefits and vicissitudes of functional knock-outs using antibodies and gene ablation. Curr. Opin. Cell Biol. 5, 786-790.

Schuch, U., Lohse, M. J., & Schachner, M. (1989). Neural cell adhesion molecule influence second messenger systems. Neuron 3, 1-20.

Simon, H., Klinz, S., Fahrig, T., & Schachner, M. (1991). Molecular association of the neural adhesion molecules L1 and N-CAM in the surface membrane of neuroblastoma cells is shown by chemical cross-linking. Eur. J. Neurosci. 3, 634-640.

Sonderegger, P., & Rathjen, F. G. (1992). Regulation of axonal growth in the vertebrate nervous system by interactions between glycoproteins belonging to two subgroups of the immunoglobulin superfamily. J. Cell Biol. 119, 1387-1394.

Tacke, R., & Goridis, C. (1991). Alternative splicing in the neural cell adhesion molecule pre-messenger RNA regulation of exon 18 skipping depends on the 5′-splice site. Genes Dev. 5, 1416-1429.

Thiery, J. P., Brackenbury, R., Rutishauser, U., & Edelman, G. M. (1977). Adhesion among neural cells of the chick embryo. J. Biol. Chem. 252, 6841-6845.

Tomasiewicz, H. Ono, K., Yee, D., Thompson, C., Goridis, C., Rutishauser, U., & Magnuson, T. (1993). Genetic deletion of a neural cell adhesion molecule variant (N-CAM-180) produces distinct defects in the central nervous system. Neuron 11, 1163-1174.

Tsiotra, P. C., Karagogeos, D., Theodorakis, K., Michaelidis, T. M., Modi, W. S., Furley, A. J., Jessell, T. M., & Papamatheakis, J. (1993). Isolation of the cDNA and chromosomal localization of the gene (Tax1) encoding the human axonal glycoprotein TAG-1. Genomics 18, 562-567.

Van Camp, B., Durie, B. G. M., Speir, C., De Waele, M., Van Riet, I., Vela, E., Frutiger, Y., Richter, L., & Grogan, T. M. (1990). Plasma cells in multiple myeloma express a natural killer cell-associated antigen: CD56 (NHK-1; Leu-19). Blood 2, 377-382.

Van Camp, G., Vits, L., Coucke, P., Lyonnet, S., Schranderstumpel, C., Darby, J., Holden, J., Munnich, A., & Willems, P. J. (1993). A duplication in the L1CAM gene associated with X-linked hydrocephalus. Nature Genetics 4, 421-425.

Volkmer, H., Hassel, B., Wolff, J. M., Frank, R., & Rathjen, F. G. (1992). Structure of the axonal surface recognition molecule neurofascin and its relationship to a neural subgroup of the immunoglobulin superfamily. J. Cell Biol. 118, 149-161.

von Bohlen und Halbach, F., Taylor, J., & Schachner, M. (1992). Cell type-specific effects of the neural cell adhesion molecules L1 and N-CAM on diverse second messenger systems. Eur. J. Neurosci. 4, 896-909.

Walsh, F. S., & Doherty, P. (1993). Factors regulating the expression and function of calcium-independent cell adhesion molecules. Curr. Opin. Cell Biol. 5, 791-796.

Walsh, F. S., Furness, J., Moore, S. E., Ashton, S., & Doherty, P. (1992). Use of the neural cell adhesion molecule VASE exon by neurons is associated with a specific down-regulation of neural cell adhesion molecule-dependent neurite outgrowth in the developing cerebellum and hippocampus. J. Neurochem. 59, 1959-1962.

Werz, W., & Schachner, M. (1988). Adhesion of neural cells to extraccellular matrix constituents. Involvement of glycosaminoglycans and cell adhesion molecules. Develop. Brain Res. 43, 225-234.

Williams, A. F. (1987). A year in the life of the immunoglobulin superfamily. Immunol. Today 8, 298-303.

Williams, E. J., Doherty, P., Turner, G., Reid, R. A., Hemperly, J. J., & Walsh, F. S. (1992). Calcium influx into neurons can solely account for cell contact-dependent neurite outgrowth stimulated by transfected L1. J. Cell Biol. 119, 883-892.

Yoshihara, Y., Oka, S., Ikeda, J., & Mori, K. (1991). Immunoglobulin superfamily molecules in the nervous system. Neurosci. Res. 10, 83-105.

Zanetta, J.-P. (1991). Is N-CAM a 'non'-cell adhesion molecule? Glycobiology 1, 317-318.

Zisch, A. H., Dalessandri, L., Ranscht, B., Falchetto, R., Winterhalter, K. H., & Vaughan, L. (1992). Neuronal cell adhesion molecule contactin/F11 binds to tenascin via its immunoglobulin-like domains. J. Cell Biol. 119, 203-213.

Zuellig, R. A., Rader, C., Schroeder, A., Kalousek, M. B., Vonbohlen, F., Osterwalder, T., Inan, C., Stoeckli, E. T., Halbach, F., Affolter, H. U., Fritz, A., Hafen, E., & Sonderegger, P. (1992). The axonally secreted cell adhesion molecule, axonin-1—Primary structure, immunoglobulin-like and fibronectin-type-III-like domains and glycosyl-phosphatidylinositol anchorage. Eur. J. Biochem. 204, 453-463.

PROTEIN ZERO OF PERIPHERAL NERVE MYELIN:

ADHESION PROPERTIES AND FUNCTIONAL MODELS

Marie T. Filbin, Donatella D'Urso, Keija Zhang, Manhar Wong, Joseph P. Doyle, and David R. Colman

Advances in Molecular and Cell Biology,
Volume 16, pages 159-192.

ISBN: 0-7623-0143-0.

I. INTRODUCTION

It is generally accepted that members of the immunoglobulin (Ig) gene superfamily of molecules are involved in adhesion/recognition (for review see Salzer and Colman, 1989). In the peripheral nervous system (PNS), several of these molecules (L1, NCAM, MAG) have been implicated in initial events in the formation of the myelin sheath. This structure is generated through interactions between single neuronal axons and individual Schwann cells (Figure 1). Early in myelinogenesis, Schwann cells individually engage a single axonal segment, flatten out against it, and begin to synthesize vast amounts of plasma membrane that wraps in a spiral around the axon. After a few turns, cytoplasm is removed and the plasma membrane becomes tightly apposed to itself. This self-attachment, or "membrane compaction" yields the characteristic, absolutely regular, periodic and lamellar arrangement of myelin. The polypeptide that mediates myelin sheath compaction in the PNS is protein zero (Po), which contains a single Ig-like segment in its extracellular domain (Lai et al., 1987; Lemke and Axel, 1985; Williams and Barclay, 1988; Lemke et al., 1988). This domain has been shown experimentally to behave as an unusually strong homophilic adhesion moiety (D'Urso et al., 1990; Filbin et al., 1990; Schneider-Schaulies et al., 1990). Po is believed to hold the extracellular surfaces of the Schwann cell's extended plasma membrane in a compact spiral around the axon, resulting in the intraperiod line. This electron-dense structure is visualized by transmission electron-microscopy, and results from the apparent fusion of the extracellular aspects of the myelin plasma membrane bilayer.

A unique feature of the Po protein, however, is that, in addition to the Ig-domain mediated interactions of its extracellular sequences, the highly charged, cytoplasmic sequences of this molecule are also believed to be adhesive and thus hold the cytoplasmic surfaces together forming a structure known as the major dense line of myelin. In addition, the interactions of the cytoplasmic domain of Po have been suggested to influence the adhesion of its extracellular domains (Wong and Filbin, 1994, 1996). The intrinsic adhesive capabilities of Po's cytoplasmic domain have yet to be demonstrated directly. A recent report, however, supports the suggestion that the cytoplasmic domain of Po interacts with acidic lipids in the opposing membrane (Ding and Brunden, 1994).

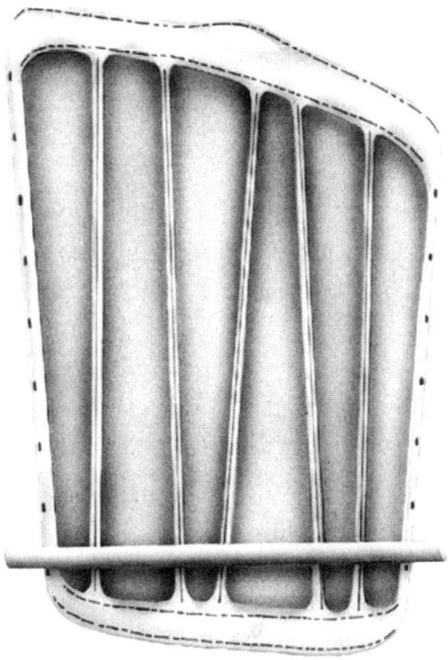

Figure 1. An opened or unwrapped Schwann cell is depicted. Compact myelin is shaded, and the cytoplasmic channels that flank the compact myelin domains are outlined.

The double adhesive role for Po in holding PNS myelin compact has long been speculated upon, largely because no other molecule is abundant enough at either the extracellular or cytoplasmic membrane surfaces to carry out this function (Kirschner and Ganser, 1980; Braun, 1984; Lemke, 1988). Compelling evidence that indeed this is the case has recently been presented from morphological examination of the peripheral nerves of mice in which no Po protein is expressed due to disruption of the Po gene through homologous recombination (Giese et al., 1992). In these mice the most frequent phenotype observed is multiple layers of loose membranes that enwrap axons but without compaction. Curiously, however, some small degree of compaction is observed in these mice; this no doubt reflects the compensatory upregulation of other adhesive proteins in these nerves.

Although Po-protein has been viewed purely as a structural molecule (Kirschner and Ganser, 1980; Braun, 1984; Lemke, 1988;

Filbin and Tennekoon, 1992), recently, however, based on observations from Po-deficient mice and on circumstantial evidence from other studies, the possibility of a more dynamic role for Po has been postulated, both in the initial stages of myelination and in possibly influencing the number of myelin lamellae that are laid down (Giese et al., 1992; Filbin and Tennekoon, 1992). These putative novel roles for Po imply involvement of signal transduction mechanisms, but no model has yet emerged describing how this could occur, nor have specific regulatory molecules or second messengers been suggested.

In a similar way, there are a number of unanswered questions regarding the recent demonstration that *in vitro* Po can promote neurite outgrowth (Schneider-Schaulies et al., 1990; Yazaki et al., 1991). Most significantly, the physiological relevance of such a function has yet to be established as current knowledge would suggest that during development a growing neurite never comes into contact with a Po molecule (Wiggins and Morell, 1980; Trapp et al., 1981). Regardless of the functional significance, because expression of Po is restricted to Schwann cells, its ability to promote neurite outgrowth can only be brought about by a heterophilic interaction with some component of the neuronal plasma membrane. The questions then arise as to what molecule Po may interact with on the neuron and, as with other adhesion molecules that promote neurite outgrowth, does a Po-neurite interaction directly transduce the signal that instructs the neurite to grow?

The adhesive properties of Po may also contribute to its sorting and accumulation in myelin. Po can be localized in cytoplasmic vesicles and it is likely that sorting is directly to the forming myelin sheath and not first to the Schwann cell plasma membrane. The signals that direct it to its final destination are unknown, but it is possible that the cytoplasmic domain of Po contains a peptide segment that acts as a "signal" that directs the molecule to myelin, perhaps, after associating transiently with a component of the cytoskeleton. It is also possible that Po is only held in the apposed myelin bilayers by interaction between its adhesive domains. That is to say, Po accumulates in myelin when the membranes are brought together and Po can interact either with itself via its extracellular domains or with some membrane component (a corresponding Po cytoplasmic domain, or perhaps a negatively charged lipid, such as phosphatidyl serine) of the opposing membrane via its cytoplasmic sequences.

In this review, we will address the traditional role of Po as a homophilic adhesion molecule, summarize what is known about the precise functional domains within its Ig domain, and how this relates to the Ig family as a whole. We will also introduce the possibility that the cytoplasmic domain of Po, in addition to being responsible for the formation of the major dense line of myelin, may also exert an influence on the interactions of the extracellular domain. Similarly, the role of the cytoplasmic domain in the sorting of Po to myelin and in signal transduction will be evaluated. Finally, the possible heterophilic interactions of Po with neurites will be discussed, as will recent suggestions that Po may be involved in certain diseases of the PNS.

II. STRUCTURE OF THE PO PROTEIN

In all mammals studied to date only a single isoform of the Po protein has been described that is translated from a single mRNA species of approximately 2.9 kb. In the chicken, however, at least five mRNA's have been described that are developmentally regulated, but the corresponding Po proteins have yet to be identified. Only the chick Po protein corresponding to the most abundant mRNA has been characterized that bears 80% sequence similarity to rat and human Po proteins (Barbu, 1990). In contrast, in the shark, based on molecular weight differences, three Po proteins have been described, but it is not known if these shark proteins result from alternative splicing of the RNA or post-translational modifications of the protein (Zand et al., 1991). The amino acid sequence for the one shark Po protein that has been published bears a 55% similarity with rat (Saavedra et al., 1989).

The rat Po cDNA was the first to be cloned from which the amino acid sequence was deduced (Lemke and Axel, 1985). Soon after, bovine Po protein was chemically sequenced (Sakamoto et al., 1987) and more recently the amino acid sequence of human Po protein was described, again deduced from its cDNA (Hayasaka et al., 1991). The amino acid sequence of all three of these mammalian Po proteins bear a 94% sequence similarity. From the amino acid sequence of the rat it was predicted that the Po protein spanned the lipid bilayer once, with 124 amino acids comprising the extracellular segment, 26 amino acids constituting the membrane spanning region and 69 amino acids exposed on the cytoplasmic surface (Lemke and Axel, 1985). This topographical distribution of domains was confirmed in biochemical studies of Po (D'Urso et al., 1990). These studies also confirmed that Po protein was synthesized with a signal sequence, which allowed insertion in the endoplasmic reticulum but which was absent from the mature Po-protein.

The most striking feature of the Po extracellular segment is the presence of a single Ig-like domain (Lai et al., 1988; Lemke and Axel, 1985; Figure 2) which encompasses virtually the entire extracellular sequence and is predicted (the crystal structure of Po has yet to be solved) to consist of nine anti-parallel β-strands that align themselves into two β-sheets, characteristic of members of this family of molecules (Williams and Barclay, 1988). The presence of nine and not seven individual β-strands sub-classifies the extracellular domain of Po as a V-like Ig domain because it more closely resembles the Ig domains found in the variable regions rather than the constant regions of immunoglobulins. Like the majority of Ig domains, the β-sheets of Po are believed to be stabilized by an intra-domain disulfide bond, which for Po is between Cys#21 in β-strand B and Cys#98 in β-strand F. The alignment of the β-strands to form two β-sheets is believed to form a common structural core for all molecules of this family which in turn allows maximum surface exposure of the amino acids in the loops between the β-sheets (Salzer and Colman, 1989; Williams and Barclay 1988). In the immunoglobulins themselves these β-sheet-intervening amino acids have been shown to carry the antigen recognition sites (Amzel and Poljak, 1979); a similar location for sites of interaction is proposed for all Ig-like domains (Williams and Barclay, 1988; Hunkapiller and Hood, 1989).

It is possible that post-translational modifications of Po can influence its conformation and/or its function (Figure 1). The Po-protein is glycosylated (Everly et al., 1973), via a single N-linkage, at Asn#93, which is contained within the Ig domain (Lemke and Axel, 1985; Sakamoto et al., 1987). Although a high-mannose type carbohydrate attachment has been reported to predominate at certain stages in development (Brunden, 1992) and in certain situations after injury (Poduslo, 1984; Brunden and Poduslo, 1987), the most abundant carbohydrate structure found on Po in mature myelin is of the complex type (Poduslo, 1984). This complex-type carbohydrate attachment, however, carries a substantial amount of

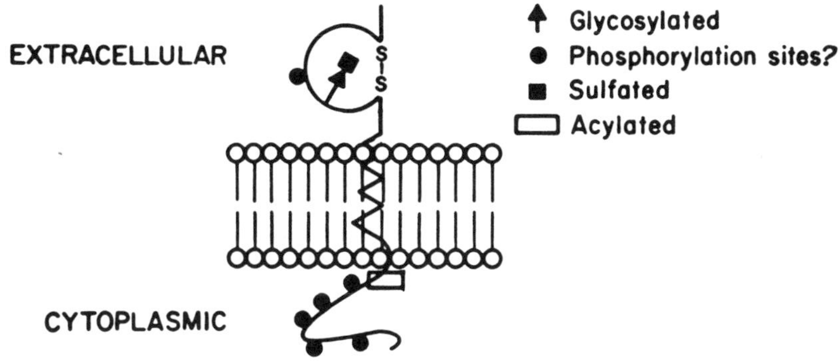

Figure 2. Schematic representation of the location of known and proposed modifications of Po-protein. The question mark in the key refers to modifications for which the precise location has yet to be determined.

microheterogeneity resulting from the presence or absence of sialic acid, galactose, and/or fucose residues (Uyemura et al., 1992). In addition, a proportion of Po molecules, which varies depending on the species, have been shown to carry the HNK 1 epitope (Bollensen and Schachner, 1987; Kuremund et al., 1988) which has been described as a sulfated glucuronic acid moiety (Chou et al., 1986). The HNK 1 epitope is carried by a number of unrelated proteins and lipids and derives its name from human natural killer cells, on which it was first identified. This sulfated glucuronic acid structure has been shown to participate directly in adhesive interactions, irrespective of the molecule to which it is attached (Abo and Balch, 1981). However, the inconsistency in the expression of this structure on Po from different species makes it an unlikely candidate to be solely responsible for the adhesive interactions of Po.

As stated previously, at 69 amino acids long, the cytoplasmic domain of Po accounts for almost a third of the molecule. Immediately following the membrane spanning region, Po is acylated at Cys#153 (Bizzozero et al., 1994), containing palmitic acid, stearic acid, and oleic acid (Agrawal et al., 1983). This single fatty acid attachment may serve to hold the cytoplasmic sequences in the required conformation for interaction. The cytoplasmic sequences carry an overall basic charge of +16, a charge comparable to that carried on myelin basic protein (MBP). Although present at low levels in PNS myelin, MBP is much more abundant in the CNS and is proposed to be Po's functional counterpart where the cytoplasmic surfaces of the CNS myelinating cell meet (Kirschner and Ganser, 1980; Braun, 1984; Lemke, 1988). Like the cytoplasmic domains of many transmembrane proteins, Po is believed to be phosphorylated on a number of residues: Ser#181, Ser#204, Ser#214 (Brunden and Poduslo, 1987; Suzuki et al., 1990). No phosphorylated tyrosine or threonine residues have been identified. Both protein kinase C and calmodulin dependent protein kinase have been implicated in the phosphorylation of Po (Brunden and Pouslo, 1987). It has been suggested that the phosphorylation of Po plays a role in the adhesive capabilities of the cytoplasmic domain. If this is the case, and because Po phosphorylation is ongoing in mature myelin, then it follows that at least a proportion of Po molecules could be undergoing adhesion and de-adhesion in compact myelin. This is not an unreasonable suggestion as the myelin membranes, like all lipid bilayers, are considered a fluid environment in which proteins have the potential to move (Braun, 1984). The extent of protein mobility within the myelin membrane and whether proteins, in particular Po, move as clusters or dimers while still interacting between opposing membranes, has yet to be determined. Alternatively, or additionally, the phosphorylation of Po could contribute to Po's suggested function in signal transduction.

III. THE ADHESION OF PO IN COMPACT MYELIN

Once Po-protein had been identified as the most abundant protein of PNS myelin it was logically proposed as the protein responsible for holding myelin compact (Kirschner and Ganser, 1980; Braun, 1984; Lemke, 1988). This proposition has been shown by a variety of approaches to be indisputably the case. Adhesion studies with Po-expressing cells (D'Urso et al., 1990; Filbin et al., 1990; Schneider-Schaulies et al., 1990; Wong and Filbin, 1994) and the analysis of mice in which the Po gene was rendered inactive through targeted homologous recombination (Giese et al., 1992) served to convincingly demonstrate this function of Po in myelin.

Although the Po-deficient mice demonstrate, as expected, the requirement for Po in the formation of compact myelin in the PNS, they give us no indication with what or how Po normally interacts to maintain the compact membrane state. Models of the adhesive interactions of both Po's extracellular and cytoplasmic domains, the first of which were presented more than 10 years ago, consisted of various combinations of homophilic and/or heterophilic interactions with either another protein or lipids of the opposing membrane (Kirschner and Ganser, 1980; Braun, 1984; Lemke, 1988). The high basic charge of Po's cytoplasmic sequences make a homophilic interaction of these domains possible, but perhaps unlikely. Similarly, the strong basic charge, as well as the low abundance of MBP on the cytoplasmic side of the myelin bilayer, may make it an unlikely partner for a heterophilic interaction with the cytoplasmic segment of Po. A transmembrane protein termed peripheral myelin protein 22 (PMP22) has recently been cloned that, although shown to be enriched in PNS myelin, is still present only at low levels (Welcher et al., 1991; Suter et al., 1992a, 1992b). The low abundance of PMP22 relative to Po, 5-7% and about 70% of the total myelin protein, respectively, make a reasonably proportional interaction of the two proteins unlikely to be responsible for holding myelin compact at the major dense line. As has been suggested, therefore (Ding and Brunden, 1994; Kischner and Ganser, 1980; Braun, 1984; Lemke, 1988), a heterophilic interaction of the cytoplasmic domain of Po with acidic lipids of the opposing membrane is a feasible model for maintaining compaction at the major dense line. Support for this model derives from the recent study that shows that the isolated cytoplasmic domain of Po can induce the rapid aggregation of artificial phospholipid vesicles (Ding and Brunden, 1994). This aggregation appears to be the result of ionic interactions and is influenced by the phosphorylation state of the cytoplasmic sequences of Po.

In contrast to the numerous interactions proposed for the cytoplasmic domain, even early models of the Po extracellular domain interactions favored a homophilic interaction (Kirschner and Ganser, 1980; Braun, 1984; Lemke, 1988). This was because the extracellular domain carried no strong charge and, again, no other protein was present at high levels in this location in myelin. Since PMP22 is a transmembrane protein, the possibility of an interaction between the extracellular domain of Po and PMP22 cannot be excluded. However, as with the cytoplasmic domain, the disparity in abundance of the two molecules makes such an interaction unlikely to account for compaction at the intraperiod line. Indeed, with the advent of recombinant DNA technology but prior to the identification of PMP22, strong support for Po as a homophilic adhesion molecule, at least via its extracellular sequences, was presented. The cloning of Po cDNA (Lemke and Axel, 1986) and the subsequent deduction of its amino acid sequence permitted the identification in the extracellular sequences of a single Ig-like domain (Lai et al., 1987). The availability of the Po cDNA allowed direct demonstration, outside the confines of the myelin membrane, of Po as a homophilic adhesion molecule via interactions of its extracellular domains. This was carried out by expressing the Po-protein after transfection of the Po cDNA in a cell line that does not usually express it, for example Chinese hamster ovary (CHO) cells (Filbin et al., 1990) or Hela cells (D'Urso et al., 1990). Through these studies it was shown, first, that where two Po-expressing cells came into contact Po accumulated at the interface of the two cells, whereas a Po-expressing cell that was not in contact with another cell had Po-protein evenly distributed over the entire cell surface (D'Urso et al., 1990). These results suggest that when Po molecules in opposing membranes come into contact with each other they interact, are thus held in that location, and accumulate. Second, by electron microscopy it was shown that at this Po-cell:Po-cell interface, the distance between the two cells was smooth and uniform and membrane thickening was frequently observed (D'Urso et al., 1990; Filbin et al., 1990; Figure 3). As similar morphological changes were not apparent when two cells not expressing Po meet, it was concluded that expression of Po induces specific cell:cell interactions. Third, in agreement with these immunolocalization studies, it was determined that cells expressing an abundance of Po-protein were at least two orders of magnitude more adhesive than the control transfected cells, that is, cells not expressing Po-protein (Filbin et al., 1990). Studies with transfected CV-1 cells and C6 glioma cells have resulted in similar findings (Schneider-Schaulies et al., 1990; Yazaki et al., 1992). In addition, by a mixed adhesion assay of Po-expressing cells and control transfected cells it was demonstrated, both qualitatively and quantitatively, that the extracellular domains of Po were interacting in a homophilic manner and not with some ubiquitous component of the opposing membrane (Filbin et al., 1990).

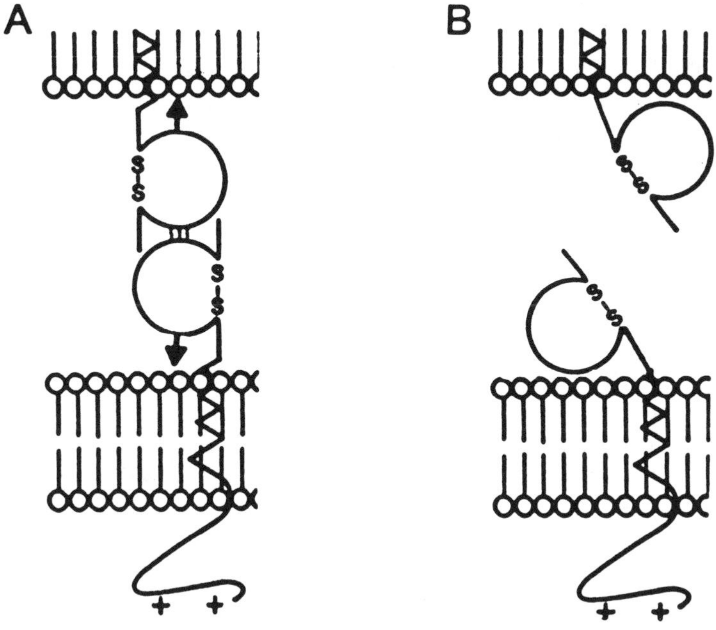

Figure 3. **(A)** Glycosylated Po is held in the correct orientation, away from the membrane from which it extends, and can interact with an opposing Po molecule. **(B)** Unglycosylated Po is not in the correct orientation. The Ig domain falls onto the membrane from which it extends and cannot interact with an opposing Po molecule.

As judged from a whole-cell re-aggregation adhesion assay, the homophilic interactions of Po are strong relative to the homophilic interactions of other cell adhesion molecules such as NCAM. That is to say, when cells in suspension, expressing an abundance of Po are permitted to aggregate, the total particle number drops by about 80% of the starting value in 40 minutes, indicating aggregate formation and strong adhesion. In contrast, under similar conditions, the total particle number of cells expressing an abundance of NCAM drops by about 10-15% in the same time (Pizzey et al., 1989). This is of interest because the strength of the interactions of adhesion molecules may play a role in their functional capabilities. It has been shown that the adhesion of NCAM that contains many sialic acid residues is weaker than NCAM that carries only a few sialic acid residues (Acheson et al., 1991). A change in polysialic acid content of NCAM has also been correlated with a change in function in that

NCAM that is heavily sialyated (and less adhesive) supports neurite outgrowth whereas NCAM that has less sialyic acid residues attached (and is more adhesive) does not (Doherty et al., 1990). Given this observation, it is possible that changes in the strength of interaction of other adhesion molecules, in particular Po, may allow for changes in function. As discussed below, it has been shown that the adhesive properties of Po change with changes in glycosylation. *In vivo*, during development and after injury, changes in the post-translational modifications of Po have been described (Brunden, 1992; Poduslo, 1984; Brunden and Poduslo, 1987). It is possible that these changes alter the function of Po, which may be required to be different at different stages in development and/or under situations of nerve injury and regeneration.

The eukaryotic cell transfection systems (D'Urso et al., 1990; Filbin et al., 1990; Schneider-Schaulies et al., 1992; Yazaki et al., 1992) that have been used to measure the homophilic binding of the extracellular domains of Po are also ideally suited for mapping functional adhesive domains in this molecule. Po-mediated adhesion has been experimentally examined in three ways: one, by mutation of the Po-cDNA prior to transfection; two, by expressing Po in different cell lines that will alter the post-translational modifications of Po, in particular the glycosylation pattern; and three, by disrupting aggregation by adding epitope-specific Po antibodies or Po-peptides to the assay. All of these approaches have been used to map the homophilic adhesive domains of Po-glycoprotein (Wong and Filbin, 1994, 1996; Yazaki et al., 1992; Griffith et al., 1992; Zhang and Filbin, 1994; Filbin and Tennekoon, 1993, 1994; Zhang et al., 1996). Because the carbohydrate composition of Po had been shown to change both with development and after axonal injury (Brunden, 1992; Poduslo, 1984; Brunden and Poduslo, 1987), the initial focus was on the role of the sugar residues in Po's homophilic adhesion. This was addressed in two ways. Initially, the effect on adhesion of changing Po from a complex to a high-mannose type glycoprotein was assessed (Filbin and Tennekoon, 1991) and, subsequently, the effect on Po's adhesiveness of eliminating the sugar residues completely was determined (Filbin and Tennekoon, 1993). The sugar residues on Po were altered from a predominantly complex-type structure to a high-mannose type structure by transfecting the Po cDNA into a glycosylation mutant CHO cell line, Lec 1A. This cell line is deficient in the enzyme N-acetylglucosaminyl transferase I, and consequently can synthesize only high-mannose type glycoproteins (Chaney and Stanley, 1986). It was found that the high-mannose form of Po was not adhesive. Hence, for Po to behave as a homophilic adhesion molecule, the sugars that distinguish complex type Po from high-mannose type Po must be present. In light of these results, it is not surprising that when the sugar residues were eliminated completely, Po no longer behaved as an adhesion molecule. However, it was also established that for adhesion to take place both molecules in the Po:Po homophilic pair must be glycosylated (Filbin and Tennekoon, 1993). This was demonstrated via a mixed adhesion assay, similar to that used in the characterization of Po as a homophilic rather than a heterophilic adhesion molecule, but this time a mixture of cells expressing either fully glycosylated Po or unglycosylated Po was used. Contrary to this finding, it has also been reported that an isolated, unglycosylated, extracellular domain of Po can bind to glycosylated Po-expressing cells grown in a monolayer (Schneider-Schaulies et al., 1990; Griffith et al., 1992). From these studies it was concluded, therefore, that only one molecule in the Po:Po homophilic pair must be glycosylated for adhesion to take place. It is possible that the interaction of unglycosylated Po with glycosylated Po is of much lower affinity than when both molecules are fully glycosylated and that the whole cell re-aggregation/adhesion assay is not sensitive enough to measure such low affinity interactions. The assay system used by the Schachner and her coworkers (Schneider-Schaulies et al., 1990; Filbin and Tennekoon, 1991), in which the ability of purified Po to bind to a monolayer of Po-expressing cells is assessed, may be a more sensitive assay and may permit the measurement of interactions of much lower affinity. An alternative explanation for this discrepancy in results is that glycosylation of the Po molecule may only be required for adhesion when Po is attached to the membrane (Figure 3). That is to say, as has been proposed by Wells and colleagues (Wells et al., 1993) from the molecular modeling of Po's extracellular domain, the sugar residues hold the Ig domain of Po away from the membrane from which it extends, in an orientation that permits amino acids of opposing Po molecules to interact; the sugar residues do not interact directly with an opposing Po molecule. In this model, if the sugars are removed the Ig domain would collapse onto the membrane and the two opposing Po molecules that normally interact would either be too far apart for, or their active sites masked from, interaction. It should be noted, however, that this model is based on the coordinates of an Ig domain from another molecule, the V_H-domain of the mouse immunoglobulin M603, as the crystal structure of Po has yet to be deduced. If, however, this model is correct, then the isolated, unglycosylated extracellular domain of Po would not be subject to this conformational constraint as it is not within the confines of the membrane. It would be free to interact with a membrane-bound, glycosylated Po molecule, held in the correct conformation. A role for the carbohydrates in maintaining the orientation of the Ig domain necessary for the adhesion of Po seems the most likely interpretation of these results, particularly in light of the results in which a specific peptide segment has been directly implicated in Po adhesion.

In vivo the Po molecule has been shown to exist as both a complex and high-mannose type glycoprotein, with the relative proportions of each changing both during development and after axonal injury (Brunden, 1992; Poduslo, 1984; Brunden and Poduslo, 1987). As described earlier, it has been shown that the adhesive properties of these two carbohydrate isoforms of Po are different. Considering these two observations together, it is possible that the *in vivo* changes in glycosylation pattern are coincident with a change in function. For example, at certain stages in development and after certain injuries, Po must change from being a "sticky" molecule to being one that does not adhere, or more precisely, one that is not required to interact homophilically. As described below, Po has recently been shown to promote neurite outgrowth (Schneider-Schaulies et al., 1990; Yazaki et al., 1991) that must involve a heterophilic interaction as neurons do not express Po. It is possible that a high-mannose form of Po favors heteropholic interactions, whereas, as has been shown, homophilic interactions only occur when a complex type Po molecule is present (Filbin and Tennekoon, 1991; Filbin and Tennekoon, 1993).

These studies on the role of the sugar residues in the adhesion of Po address only two extremes: when the sugars that comprise a complex type structure have been eliminated, leaving only a high-mannose structure, and when all the sugars have been removed. *In vivo*, in addition to the co-existence of complex and high-mannose type Po, considerable microheterogeneity has been reported for the complex-type structure with differences between molecules of only one sugar moiety such as sialic acid, galactose or fucose. In addition, a portion of Po, in some species, carries the HNK 1 epitope (Bollensen and Schachner, 1987; Kutemund et al., 1988). Although this microheterogeneity may appear minor, it should be noted that the presence or absence of one sugar residue such as sialic acid can change the charge of Po and thus perhaps affect its interactions. It is not known if or how these minor variations in the sugar composition of Po affect its conformation/orientation, which in turn could affect the exposure of different functional domains.

The role of sugar residues in the functioning of other adhesion molecules in the Ig gene superfamily is not as readily addressed as it is for Po, because most other members of this family of molecules have multiple Ig domains and/or multiple glycosylation sites. Even so, the sialic acid content of NCAM has been implicated in altering its adhesive properties (Acheson et al., 1991) and the switch from a complex to a high-mannose type sugar structure for L1 impairs its ability to interact in a *cis* fashion with NCAM (Kadmon et al., 1990). It is not known which of the many glycosylation sites on these molecules is responsible for these changes in function. This is not the case for the human immune system molecule, CD2, where elimination of a single glycosylation site from Ig domain 1 completely abolishes its adhesive properties (Recny et al., 1992). This glycosylation site is the only site of sugar attachment on Ig domain 1 of CD2, which, of its two Ig domains, has been shown to be responsible for adhesion. Similar to Po, it is suggested that the sugars of CD2 do not themselves interact with an opposing molecule but instead hold this Ig domain in an orientation optimal for adhesion.

If, as is suggested above, the sugar residues of Po are not directly involved in the interactions that hold two Po molecules together, it follows that this function must be carried out by amino acids of the extracellular domain. Since all the amino acids of Po's extracellular sequences are part of the Ig domain, the prediction of which are exposed at the surface of the folded molecule and therefore in a position capable of interaction, can be based on the alignment of the β-strands proposed for all V-like Ig domains (Williams and Barclay, 1988). This alignment, for the majority of molecules, is generally believed to be stabilized by a disulfide bond between two cysteine residues 50-70 amino acids apart. Hence, the formation of this disulfide bond is directly responsible for the stability of the functional conformation of Ig domains. As was the case for addressing the role of the sugar residues in adhesion, Po is the ideal molecule upon which to test this hypothesis as it has only one Ig domain and hence a single disulfide bond. The formation of the Po Cys#21-Cys#98 disulfide bond was prevented by changing Cys#21 to an alanine by mutation of the Po cDNA. When expressed in CHO cells, the mutated Po protein still reached the cell surface and was glycosylated but did not behave as an adhesion molecule (Zhang and Filbin, 1994). Therefore, as has long been predicted for all members of the Ig superfamily, the function of Po-protein, which may represent the "prototypic" Ig family member, is dependent on the formation of the characteristic disulfide bond within the Ig domain.

The necessity of an intact Ig-domain disulfide bond for adhesion of Po strengthens the hypothesis that this bond stabilizes the alignment of the hydrophobic β-strands, thus stabilizing the exposure of the amino acids in the loops between the β-strands at the surface of Ig molecules. Hence, these amino acids are most likely to participate in binding to other molecules. It was found, by two groups (Yazaki et al., 1992; Zhang et al., 1996), using peptides and peptide antibodies corresponding to these intervening sequences in Po's Ig domain, that a five amino acid sequence, SDNGT (from amino acid#91-#95), is directly involved in Po's adhesion: both SDNGT-peptide antibodies and the peptide itself block adhesion completely. Furthermore, when ASP #92 and Gly #94 in this sequence mutated, adhesion is lost (Zhang et al., 1996). The SDNGT sequence lies between β-strand E and β-strand F, and is of particular interest for a number of reasons: it spans the glycosylation site at Asn#93; two amino acids at Asp#92 and at Gly#94 are highly conserved at this position in a large number of V-like Ig domains; and, finally, this SDNGT sequence is found in its entirety, in the haemagglutinin protein of the 1976 strain of the influenza virus, which when used as a vaccine resulted in an outbreak of the demyelinating disease, Guillian-Barre syndrome (GBS) (Schonberger et al., 1979; Safranek et al., 1991). Because this sequence includes the glycosylation site and because the sugar residues are necessary for adhesion of Po it could be suggested that this SDNGT sequence interacts with the sugar residues within the same molecule, and in so doing holds the Ig domain in the correct conformation. An alternative interpretation of the data is that these amino acids interact directly with either the same or some other amino acid sequence on an opposing Po molecule. It is of interest that Po-peptide #74-82 could inhibit Po:Po adhesion by about 80% which raises the possibility that Po amino acids #74-82 on one Po molecule interact with Po amino acids #91-95 on an opposing Po molecule (Zhang et al., 1996).

Conservation of two of the SDNGT amino acids, Asp#92 and Gly#94, at this location in a majority of V-like Ig domains (Williams and Barclay, 1988) suggests that they have some functional relevance in this sub-set of Ig family members. In addition, as this sequence is found, with at most one conservative substitution, in a number of other Ig-like molecules it may represent a sequence that is a common Ig-family adhesive motif (Horlick et al., 1992; Newman et al., 1990; Yarden et al., 1986). To establish if this is the case, the effect of mutating this sequence on the adhesive abilities of not only Po but of other molecules that carry this sequence, needs to be determined. In addition, if insertion of the SDNGT sequence into a molecule that is not normally adhesive confers the molecule with adhesive properties, then this sequence could be regarded with more confidence as an adhesive motif.

As referred to in the introduction, the cytoplasmic domain of Po is believed to be responsible for holding the myelin membranes together at the major dense line. It is also possible, as with other adhesion molecules such as the cadherins (Nagafuchi and Takeichi, 1988) and the integrins (Hynes, 1992), that the cytoplasmic domain of Po influences the interactions of the extracellular sequences. Although the two are not necessarily exclusive events, the putative interactions of the cytoplasmic domain of Po in holding myelin together at the major dense line will be discussed below.

Recently, it was established that indeed the cytoplasmic domain of Po is required for the adhesion of its extracellular domain (Wong and Filbin, 1994, 1996). This was determined by assessing the ability of two truncated Po proteins to behave as homophilic adhesion molecules. The truncations resulted in Po-proteins lacking either the last 59 amino acids or the last 52 amino acids from the cytoplasmic sequences. Both of the truncated Po-proteins, when expressed in CHO cells after transfection of truncated Po cDNA's, reached the cell surface and were glycosylated to the same extent as the full-length protein. Neither of the truncated Po-proteins, however, behaved as homophilic adhesion molecules, which demonstrates that the cytoplasmic sequences of Po must be intact for adhesion of the extracellular domains to take place. How could this be brought about when a lipid bilayer separates the two domains? It is possible that the initial interactions of the extracellular domains induce a conformational change in the cytoplasmic domain which in turn serves to stabilize and strengthen the extracellular interactions. Such a model involves a two-step adhesive interaction whereby the first step, the initial interaction of the extracellular domains, is of low affinity and is independent of the cytoplasmic sequences. The second step is of much higher affinity and is dependent on the presence of the cytoplasmic domain. Such a two-step adhesive interaction has also been proposed to explain why truncated forms of the cadherins when expressed in cells do not adhere, whereas an isolated extracellular domain of N cadherin is capable of inhibiting cadherin:cadherin mediated cell adhesion (Nagafuchi and Takeichi, 1988). It is proposed that the initial, low affinity adhesion is not strong enough to hold two cadherin-expressing cells together but is of sufficient strength to permit the isolated extracellular domain to competitively inhibit the interaction of cadherins on opposing cells. If this model is correct, for either Po or cadherins, the next question posed is what interactions of the cytoplasmic sequences could increase the adhesiveness of the extracellular domain? Cadherins have been shown to interact, via another protein, with the cytoskeleton and it is this interaction that is believed to strengthen the interactions of the extracellular domains (Nagafuchi et al., 1991; Hirano et al., 1992).

For another family of adhesion molecules, the integrins, clustering within the plane of the membrane is required for lymphocyte activation (Hynes, 1992; Zachary and Rosengunt, 1992). In this situation, integrin clustering again via cytoplasmic domain interactions with the cytoskeleton, also recruits non-receptor tyrosine kinases to the membrane which in turn is believed to be responsible for transducing the activating signal (Huang et al., 1993). It is most likely that integrin clustering strengthens the cell:cell or cell:extracellular matrix, integrin-mediated adhesion of these cells. In view of the precedent set for both cadherins and integrins, it is not unreasonable to suggest that the adhesive interactions of the extracellular domains of Po are strengthened, by an association of the cytoplasmic domains with the cytoskeleton, which in turn may or may not induce clustering of Po molecules. An association of Po with the cytoskeleton seems likely based on two observations. First, the estimation that approximately 30% of the protein, either expressed by a Schwann cell line or in transfected CHO cells, is insoluble in the detergent NP40. In contrast, only 5-10% of Po lacking the last 52 amino acids of the C-terminal is insoluble while Po lacking 59 amino acids, under similar conditions, is completely soluble

(Wong and Filbin, 1994). This non-ionic detergent will solubilize integral membrane proteins but not cytoskeleton proteins nor their associated proteins. Second, the microtubule disrupting agent, colchicine, abolishes binding of full-length Po (Wong and Filbin, 1996). If these results are considered in light of the two-step model of Po adhesion just described, the first, low affinity step would induce binding of the cytoplasmic sequences with another protein, possibly a component of the cytoskeleton (Figure 4). Such an association would in turn influence the behavior of the extracellular domains, perhaps by inducing clustering, so that their adhesion is strengthened. As there is no cytoskeleton in compact myelin the Po/cytoskeleton association must be transient. The sequence of events described above would take place at the early stages of myelination, as the membranes are coming together and before cytoplasm and cytoskeletal elements have been extruded from the leaflets (Figure 4). Electron microscopic examination of peripheral nerve during the early stages of myelination suggests that indeed the extracellular domains of the myelin membrane come together and adhere before the cytoplasm is extruded and the cytoplasmic sequences of Po meet (Peter et al., 1991). Upon compaction, the interactions of the cytoskeleton with Po's cytoplasmic domain would be replaced by interactions with a component of the opposing membrane and consequently, any influence cytoskeletal interactions have on the adhesiveness of the Po extracellular domains would be transferred (taken over) by this novel interaction. An active role for the cytoskeleton in bringing the myelin membranes together for compaction can also be incorporated into this model. That is to say, if the cytoskeleton rearranges and pulls back towards the nucleus, because of its interaction with the cytoplasmic domain of Po, this would bring the cytoplasmic surfaces together and permit components of the membranes to interact (Figure 4).

The question of dimerization or clustering via *cis* interactions of Po's cytoplasmic domains, either directly or via the cytoskeleton, as a prerequisite for strong *trans* interactions of the extracellular domains has been addressed in two ways, by producing what could be loosely termed dominant/negative mutant cell lines for Po function and by chemical cross-linking studies (Wong and Filbin, 1996). For the dominant/negative studies, it was reported that when Po, truncated in the cytoplasmic domain, is coexpressed with full-length Po, full-length Po does not ahdere. In keeping with this

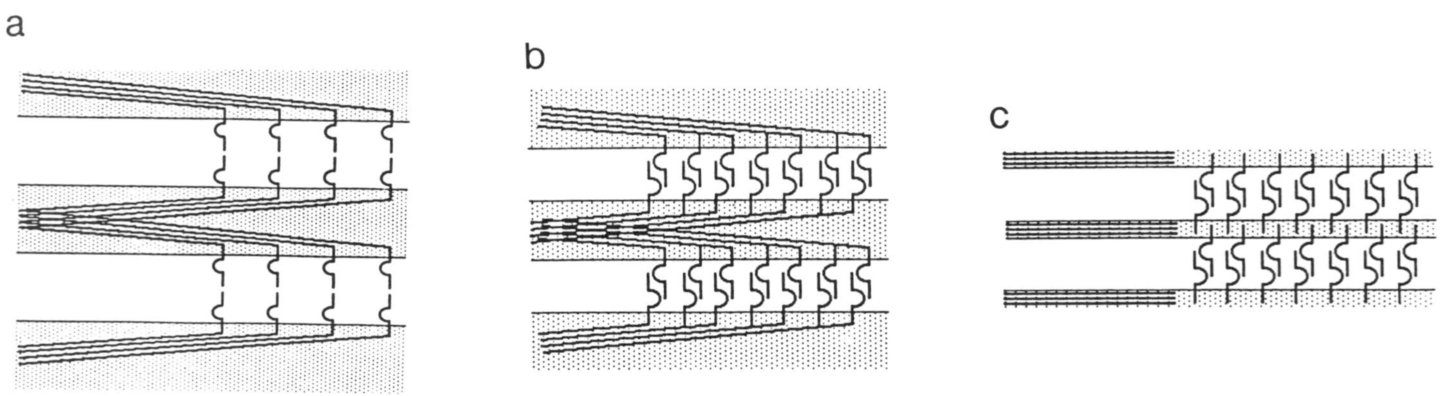

Figure 4. Model of the Interactions of Po During the Compaction of Myelin. (**A**) An initial low affinity interaction of Po's extracellular domains (represented by half-circles in the clear region) triggers a change inn the interaction of the cytoplasmic domain (represented by the stippled areas) with the cytoskeleton. (**B**) The cytoskeleton reorganizes and pulls back towards the nucleus, inducing Po clustering which in turn strengthens the adhesion of the extracellular domains. (**C**) The cytoskeleton pulls back until the cytoplasmic domains of Po are brought into contact with an opposing membrane. The cytoskeleton disengages. Myelin becomes compact.

observation, after chemical cross-linking and immunoprecipitation, it was concluded that when expressed alone full-length Po clusters and forms large aggregates but when coexpressed with truncated Po, no aggregates are apparent. This suggests that the presence of the truncated Po is preventing cluster formation, which in turn is preventing adhesion (Wong and Filbin, 1996). In the long term, similar experiments could be conducted with transgenic mice and the ability of a Po-transgene, missing the cytoplasmic domain, to disrupt myelination assessed.

Clustering of Po as a prerequisite for *trans* adhesion has been suggested by Griffith and colleagues (1992) but the clusters are suggested to form through *cis* interactions of the extracellular domains. No involvement of the cytoplasmic domains is proposed. In this model the carbohydrate, HNK-1 moiety of one Po molecule is proposed to interact with amino acids of a Po molecule in the same membrane. These studies were conducted with Po isolated from human sciatic nerve, a high proportion of which carries the HNK-1 epitope. This, however, is not the case for Po from other species; little or no HNK-1 can be detected on Po isolated from rat (O'Shannessy et al., 1985). Consequently, extracellular domain Po:Po *cis* interactions, which are dependent on the HNK-1 epitope, cannot apply to all species. It should be noted, however, that in all species studied to date that PMP22 has been shown to carry the HNK-1 epitope. The ability of Po to interact with PMP22 remains to be tested.

Alternatively, the putative interactions of the Po cytoplasmic domain with the cytoskeleton may have no bearing on the interactions of the extracellular domain, but instead may be important only for the sorting of Po to myelin. Immunolocalization studies of Po in nerves in which active myelination is occurring suggest that Po is sorted, directly, in vesicles to the myelin membranes (Trapp et al., 1992), as Po can only be detected on the Schwann cell surface in the absence of axonal contact, that is, when no myelin is present (Trapp et al., 1992, 1993; Mirsky et al., 1980; Rutkowski et al., 1990). This observation in turn suggests that different mechanisms of Po sorting are in place under the two situations; one that sorts Po to myelin and one that sorts Po to the surface, perhaps as a default pathway. As it has been shown that a truncated Po protein, missing the last 59 amino acids from the cytoplasmic domain, still reaches the cell surface in transfected CHO cells but is completely soluble in NP40 (Wong and Filbin, 1994), it is unlikely that an interaction of Po with the cytoskeleton is required for sorting to the cell surface. On the other hand, because Po has been detected in vesicles in sections of peripheral nerve (Trapp et al., 1993), it is possible that these vesicles are sorted directly to myelin via an interaction of Po's cytoplasmic domain, extruding from the vesicle, with the cytoskeleton. The expression of Po in the Schwann cell plasma membrane may be the result of a default pathway involving a constitutive sorting mechanism when myelin, the destination for the active sorting of Po, is not present. Once Po reaches myelin it could be held there by the adhesive interactions of both its extracellular domains and its cytoplasmic domains.

Regardless of whether the cytoplasmic sequences of Po-protein influence the adhesion of the extracellular domains, these sequences are believed to be responsible for maintaining the myelin membranes compact at the major dense line. The most compelling evidence in support of such a role for these Po sequences comes from studies on the mutant mouse, the *shiverer*. The mutation in this mouse has been identified as a deletion in the gene that codes for MBP (Roach et al., 1983). Myelin basic protein is expressed in both the CNS and the PNS although the levels of expression are very different in the two systems; 30-40% of the total myelin protein of CNS myelin but only 5-15% of PNS myelin protein. In the *shiverer* mutant mouse no MBP is expressed and, in the CNS, hypomyelination is apparent. The little compact membrane that is present does not have a major dense line. On the other hand, the PNS myelin of these mice appears to be normal. As Po is the only protein found in any abundance at the major dense line in the PNS, it is concluded that it holds together the myelin membranes at this location rendering the low levels of MBP unnecessary for this function in the PNS (Kirschner and Ganser, 1980). The role of PMP22 in the maintenance of this structure has yet to be determined, but the negative charge carried by the predicted cytoplasmic sequences of PMP22 would complement the basic charge of Po. Alternatively, it has been proposed that Po holds the membranes together by interacting with acidic lipids in the opposing membrane (Braun, 1984; Lemke, 1988). In support of such an interaction, recently, the isolated cytoplasmic domain of Po has been shown to induce the aggregation of artifical phospholipid vesicles (Ding and Brunden, 1994). This aggregation was shown to be affected by changes in the ionic charge of either

Po (by altering the degree of phosphorylation) or of the lipid membrane (by decreasing the phosphatidylserine content).

IV. PO-MEDIATED ADHESION AND NEURITE OUTGROWTH

Like numerous other members of the Ig family of molecules, Po has recently been shown to promote neurite outgrowth in culture (Schneider-Schaulies et al., 1990; Yazaki et al., 1991). While there is no doubt that Po can induce such a phenomenon *in vitro*, the physiological relevance of this capability is questionable. This is because it is widely believed that a growing neurite never comes into contact with a Po molecule (Wiggins and Morell, 1980; Trapp et al., 1981, 1993). Until recently, Po was believed to be expressed only after the Schwann cell had segregated single, large axons that were destined, because of their size, to be myelinated. At this stage in development the axon has already reached its target and the growing tip has long since traversed the section of nerve being ensheathed by the Schwann cell. Data from the chick places the initial expression of Po at a much earlier time point, even before Schwann cells have fully differentiated, while they are still migrating as part of the neural crest (Bhattacharyya et al., 1991; Sherman et al., 1993). In the rat, Po has been shown to be expressed as early as when the Schwann cell first makes contact with the axon (Martini et al., 1988; Lamperth et al., 1991). This, however, would still be too late to have any influence on neurite outgrowth. Based on results from the chick, it is possible that low, previously undetected, levels of Po are expressed very early in the rat at a time when axonal outgrowth is taking place. Alternatively, rather than influencing neurite outgrowth during development, Po may play a role in nerve regeneration after injury when the regenerating axon tip would most certainly come into contact with Schwann cells and myelin debris expressing Po.

Irrespective of the functional relevance of Po's ability to promote neurite outgrowth, its interaction with the neuron, in culture or otherwise, must be the result of a heterophilic interaction because Po is only expressed by Schwann cells. Other Ig family members that induce neurite outgrowth, such as NCAM and L1 (Doherty and Walsh, 1992), do so via a homophilic interaction, in which case the same molecule is acting as receptor on one cell and as ligand on the other. In the case of Po, it is behaving as a ligand and the receptor with which it interacts on the neuronal cell remains to be identified. The studies in which Po has been shown to promote neurite outgrowth have used only dorsal root ganglia neurons which, *in vivo*, project axons into the peripheral nervous system. It would be of interest to determine if other neurons, say cerebellar neurons from the CNS, respond in a similar way to Po. This would give an indication of whether the Po receptor is confined to PNS neurons or if it is expressed more ubiquitously in the nervous system. If the latter is found to be the case it is possible that Po represents a generic structure, which may interact not with a specific receptor but with many different molecules on different neurons, which when activated/bound promote neurite outgrowth. As Po is the smallest and simplest member of the Ig family of molecules and thought to represent the closest relative to the ancestral gene for the whole family (Williams and Barclay, 1988; Lemke et al., 1988), this is not an unreasonable prediction.

To date, only the functional domains involved in Po homophilic binding have been mapped. It will be of interest to determine if the same domains are responsible for Po heterophilic interactions, and to determine if a switch in function (from homophilic adhesion in compact myelin to heterophilic adhesion in neurite outgrowth) is accompanied by a switch in structural requirements.

V. PO-PROTEIN AND SIGNAL TRANSDUCTION

It is possible that Po may be a player in a signal transduction cascade, although direct evidence for this has not been forthcoming. In support of the idea, as outlined above, it has recently been demonstrated that, contrary to previously held notions, Po is expressed at the very early stages of myelination (Bhattacharyya et al., 1991; Sherman et al., 1993; Martini et al., 1988; Lamperth et al., 1991). At this time in myelination many proteins are up-regulated while others are dramatically down-regulated. The orchestration of these events must involve signal transduction and Po may in fact play some role in the regulation of expression of these molecules. Additionally, in the transgenic mice just described in which the Po gene has been "knocked-out" by homologous recombination and in which no Po protein is expressed, there is a mis-regulation of numerous proteins compared to myelination in normal mice; NCAM and MAG are not appropriately down-regulated, while MBP is not up-regulated (Giese et al., 1992). The question of the role of Po in these events, as is the situation for a number of other adhesion molecules in similar situations, is whether this molecule is directly responsible for transducing the signal that initiates these events. Alternatively, is it just the bringing together of two cells (or membranes) which in turn permits the interaction of other molecules capable of initiating the signal? It is possible that, while not necessarily transducing the signal directly, adhesion molecules, in addition to their homophilic interactions, specifically interact with an adaptor molecule which transduces the signal (Doherty and Walsh, 1994). On the other hand, the fact that a sequence motif in the cytoplasmic domain of Po has, when present in other proteins, been reported to specifically bind G-proteins (Nishimoto et al., 1993) supports a direct role for Po in a signal transduction mechanism. However, the binding of G-proteins to Po has not yet been demonstrated.

A role for Po in signaling could also be envisaged in the later stages of myelination in that the final number of myelin lamellae which wrap the axon is directly proportionate to the diameter of the axon. As Po is the most abundant protein of PNS myelin, and as it spans the membrane and is proposed to interact both via its extracellular and via its cytoplasmic domains, it could perhaps be responsible for informing the nucleus of the Schwann cell that enough membranes have been laid down. Is it possible for a signal to be passed through the layers of myelin? Again, the ongoing phosphorylation and dephosphorylation of Po in mature myelin (Brunden and Poduslo, 1987) lends credence to such a hypothesis. This notion is supported further by the following observations in mice bred to mis-express Po and/or MBP. Mice deficient in MBP but with normal levels of Po or mice with normal levels of MBP but heterozygote for Po deficiency (half the normal amount of Po) each have the usual number of myelin lamellae surrounding axons. In contrast, mice heterozygote for Po, but which are devoid of MBP, have a reduced thickness of myelin around normal axons (Martini et al., 1995a). This suggests that both Po and MBP play a role in determining myelin thickness.

VI. IMPLICATIONS OF THE ADHESION OF PO TO THE IG SUPERFAMILY

As previously stated, Po has been proposed to be the closest relative to the ancestral gene for the entire Ig superfamily of molecules (Willilams and Barclay, 1988; Lemke et al., 1988). What exactly this means and what relevance it has to the common functioning of these molecules remains to be precisely defined. The prevailing hypothesis is that the original, ancestral Ig-like molecule was a very simple molecule, containing at most only a single Ig-like domain. It is proposed that duplication of this single domain resulted in the evolution of all other family members, the majority of which contain multiple Ig-like domains making them more complex but also potentially increasing their specificity (Williams and Barclay, 1988; Hunkapiller and Hood, 1989). Indeed, the sequences coding for the single Ig domain of Po are divided almost symmetrically between two exons (Lemke et al., 1988). This in turn suggests that the original Ig molecule may have corresponded to what is now regarded as half an Ig-domain and the full Ig-domain resulted from duplication of this half domain. It is because of the exon distribution of the Po Ig domain sequences and because of the size and simplicity of the molecule that Po is touted as being the closest relative to the ancestral gene for the whole family of molecules. If this is so, does this relationship bear any functional significance to the entire family of molecules? Is Po the prototypic adhesion molecule for the Ig superfamily? Are there generic sequences/structures necessary for adhesion that are carried by Po which have been conserved throughout evolution? Are these sequences/structures common to all adhesion molecules of the Ig superfamily? The SDNGT sequence identified as being directly involved in Po:Po adhesion (Yazaki et al., 1992; Zhang et al., 1996) may represent such a generic, adhesion sequence, at least for a sub-set of Ig-family members, V-like Ig domains.

However, if Po is the closest relative to the Ig ancestral molecule, it is surprising that it appears relatively late in evolution in elasmobranchs (sharks). Curiously, no Po-like molecule has been identified in any cell other than the Schwann cell. Other molecules with Ig-like domains have been identified in more primitive animals, including yeast (Doherty and Walsh, 1994). It is therefore perhaps more likely that Po arose by gene convergence; that is, it is a relatively new molecule in the vertebrate repertoire that arose with the invention of the myelin sheath in vertebrates.

VII. DISEASES INVOLVING PO PROTEIN

Until very recently none of the peripheral dysmyelinating human diseases and none of the well-characterized dysmyelinating mutant mice were associated with a defect in Po. While there is still no dysmyelinating animal model resulting from abnormalities in Po, a sub-set of patients suffering from the dysmyelinating disease, Charcot-Marie-Tooth type 1B (CMT 1B), have been shown to have point mutations in the Po gene (Kulkens et al., 1993; Hayasaka et al., 1993). To date seven different point mutations or single acid deletions have been identified. Except for one of these mutations, which is in the transmembrane domain, all are in the extracellular sequences (for review see Patel and Lupski, 1994). Defects in Po protein could bring about the dysmyelinating phenotype by a number of different mechanisms such as abberent insertion of the protein into the membrane, instability of the protein in the membrane, and/or loss of adhesive function. Morphological examination of the sural nerve from two patients from one of the pedigrees revealed that there was a little compact myelin present in these individuals and that this myelin had normal spacing. On the other hand, there was a tremendous reduction in the number of myelinated axons and the myelin sheaths that were apparent were very thin. In addition, an abundance of tomacula were reported that represent what is believed to be redundant myelin sheaths, that is, not wrapping an axon. Myelin-like debris was also observed within the cytoplasm of Schwann cells. All of these morphological abnormalities are associated with demyelination. Similarly, frequent "onion bulb" formations were observed, which are indicative of attempted remyelination. Finally, from immunostaining of tissue sections, the level of expression of Po protein was judged to be normal in these patients (Thomas et al., 1993). It is of particular interest that in all of the pedigrees of CMT 1B associated with a mutation in Po, the individuals affected are heterozygotes for the defective gene. As a consequence 50% of the Po protein produced by these individuals is presumed normal. Recently it has been reported that myelin in heterozygote Po-deficient mice is apparently normal until the mice are about four months old at which time demyelination occurs (Martini et al., 1995b). While this pattern of disease mimics some of the CMT 1B milder phenotypes, reduced levels of Po cannot explain the more severe phenotypes which have an earlier onset. The presence of the mutated Po protein could be directly responsible for the abnormalities observed in this disease. How this is brought about has yet to be determined, but could be explained if the mutated Po were having a dominant/negative effect on the function of the wildtype Po protein. We have suggested that Po must cluster within the plane of the membrane in order to interact effectively with Po in the opposing membrane. It is possible that if a number of Po molecules, which as a consequence of point mutations in the extracellular domain will not adhere, are included in any one cluster, there may not be enough functional Po molecules within that cluster to interact stabily with an opposing membrane. This in turn could result in the transient formation of compact myelin, demyelination, and subsequent attempts to remyelinate. Interestingly, a similar morphological phenotype is observed with patients suffering from CMT 1A, which is associated with either a point mutation or a duplication of PMP22 (Suter et al., 1992). Perhaps, as suggested above, an as yet unidentified association of Po and PMP22 that is essential to the formation of myelin is disrupted in both CMT1A and CMT 1B. This would account for a similar phenotype resulting from mutations in two separate genes.

The Po protein may also play a role in the etiology of the demyelinating disease, Guillain-Barre syndrome (GBS), at least in a sub-set of patients. An outbreak of GBS has been directly correlated to vaccination with the New Jersey strain of swine influenza virus in 1976 (Schonberger et al., 1979; Safranek et al., 1991). This particular strain of the virus, and not other strains which, when used as a vaccine did not result in an outbreak in GBS, contains the SDNGT sequence, identified as being directly involved in Po:Po adhesion. It is possible that vaccination with the SDNGT containing influenza virus resulted in the production of antibodies to this sequence. These antibodies could then cross-react with the intact Po protein in myelin and in so doing initiate demyelination. It remains to be determined if these patients have antibodies to Po. However, the demyelinating disease, experimental allergic neuropathy (EAN), has been induced in animals by immunization with Po protein (Milner et al., 1987).

VIII. SUMMARY

In this review we have described the homophilic and the heterophilic *trans* (with itself or with a component of the neuronal plasma membrane) and *cis* (with itself) interactions of Po's extracellular domain. We have also outlined the possible interactions of the cytoplasmic domain in a *cis* (with itself) or *trans* (with acidic lipids) interaction and with the cytoskeleton. Also, we have suggested how the interactions of the extracellular domain and the cytoplasmic domain may influence each other. Finally, we have speculated on the involvement of Po in signal transduction and in human diseases. For such a simple molecule the complexities of its interactions are continually expanding.

REFERENCES

Abo, T., & Balch, C.M.A. (1981). Differentiation antigen of human NK and K cells identified by a monoclonal antibody (HNK-1). J. Immunol. 127, 1024-1029.

Acheson, A., Sunshine, J.L., & Rutishauser, U. (1991). N-CAM polysialic acid can regulate both cell-cell and cell-substrate interactions. J. Cel. Biol. 114, 143-153.

Agrawal, H.C., Schmidt, R.E., & Agrawal, D. (1983). *In vivo* incorporation of H^3 Palmitic acid into Po protein, the major intrinsic protein of rat sciatic nerve. J. Biol. Chem. 258, 6556-6560.

Amzel, L.M., & Poljak, R.J. (1979). The three-dimensional structure of immunoglobulins. Annu. Rev. Biochem. 48, 961-997.

Barbu, M. (1990). Molecular cloning of cDNA's that encode the chicken Po protein: Evidence for early expression in avians. J. Neurosci. Res. 25, 143-151.

Bhattacharyya, A., Frank, E., Ratner, N., & Brackenbury, R. (1991). Po is an early marker of the Schwann cell lineage in chickens. Nueron 7, 831-844.

Bizzozero, O.A., Fridal, K., & Pastuszyn, A. (1994). Identification of the palmitoylation site in rat myelin Po glycoprotein. J. Neurochem. 62, 1163-1171.

Bollensen, E., & Schachner, M. (1987). The peripheral myelin glycoprotein Po expresses the L2/HNK-1 and L3 carbohydrate structures shared by neural adhesion molecules. Neurosci. Lett. 82, 77-82.

Braun, P.E. (1984). Molecular organization of myelin. In: *Myelin* (P. Morell, Ed.). Plenum Press, New York, pp. 97-116.

Brunden, K.R. (1992). Age dependent changes in the oligosacchoride structure of the major myelin glycoprotein, Po. J. Neurochem. 58, 1659-1666.

Brunden, K.R., & Poduslo, J.F. (1987). Lysosomal delivery of the major myelin glycoprotein in the absence of myelin assembly: Posttranslational regulation of the level of expression by Schwann cells. J. Cell. Bio. 104, 661-669.

Brunden, K.R., & Poduslo, J.F. (1987). A phorbol ester-sensitive kinase catalyzes the phosphorylation of Po glycoprotein in myelin. J. Neurochem. 46, 1863-1872.

Bunge, R.P., Bunge, M.B., & Bates, M. (1989). Movements of the Schwann cells nucleus implicated progression of the inner (axon-related) Schwann cell process during myelination. J. Cell Biol. 109, 273-284.

Chaney, W., & Stanley, P. (1986). Lec1A chinese hamster ovary cell mutants appear to arise from a structural alteration in N-acetylglucosaminyltransferase I. J. Biol. Chem. 261, 10551-10557.

Chou, D.K.H., Ilyas, A.A., Evans, J.E., Costello, C., Quarles, R.H., & Jungalwala, F.B. (1986). Structure of sulfated glucuronyl glycolipids in the nervous system reacting with HNK-1 antibody and some IgM paraproteins in neuropathy. J. Biol. Chem. 261, 11717-11725.

Ding, Y., & Brunden, K.R. (1994). The cytoplasmic domain of myelin glycoprotein Po interacts with negatively-charged phospholipid bilayers. J. Biol. Chem.

Doherty, P., Cohen, J., & Walsh, F.S. (1990). Neurite outgrowth in response to transfected N-CAM changes during development and is modulated by polysialic acid. Neuron 5, 209-219.

Doherty, P. & Walsh, F.S. (1992). Cell adhesion molecules, second messengers, and axonal growth. Curr. Opin. Neurobiol. 2, 595-606.

Doherty, P. & Walsh, F.S. (1994). Signal transduction events underlying neurite outgrowth stimulated by cell adhesion molecules. Curr. Opin. Neurobiol. 4, 49-55.

Doyle, J.P., Stemack, J.G., Cowin, P., Colman, D.R., & D'Urso, D. (1995). Protein zero, a nervous system adhesion molecule, triggers epithelial reversion in host carcinoma cells. J. Cell Biol. 131, 465-482.

D'Urso, D., Brophy, P.J., Staugaitus, S.M., Gillespie, C.S., Frey, A.B., Stempak, J.G., & Colman, D.R. (1990). Protein zero of peripheral nerve myelin: Biosynthesis, membrane insertion and evidence for homotypic interaction. Neuron 2, 449-460.

Everly, J.L., Brady, R.O., & Quarles, R.H. (1973). Evidence that the major protein in rat sciatic nerve myelin is a glycoprotein. J. Neurochem. 21, 329-334.

Filbin, M.T., & Tennekoon, G.I. (1991). The role of complex carbohydrates in adhesion of the myelin protein, P. Neuron 7, 845-855.

Filbin, M.T., & Tennekoon, G.I. (1992). Myelin Po-protein, more than just a structural protein? Bioessays 14, 541-547.

Filbin, M.T., & Tennekoon, G.I. (1993). Both Po-molecules must by glycosylated for homophilic adhesion. J. Cell Biol. 122, 451-459.

Filbin, M.T., Walsh, F.S., Trapp, B.D., Pizzey, J.A., & Tennekoon, G.I. (1990). Role of myelin Po-protein as a homophilic adhesion molecule. Nature 344, 871-872.

Giese, K.P., Martini, R., Lemke, G., Sariano, P., & Schachner, M. (1992). Mouse Po gene disruption leads to hypomyelination, abnormal expression of recognition molecules and degeneration of myelin and axons. Cell 71, 565-576.

Griffith, L.S., Schnitz, B., & Schachner, M. (1992). L2/HNK-1 Carbohydrate and protein-protein interactions mediate the homophilic binding of the neuron adhesion molecule Po. J. Neurosci. Res. 33, 639-647.

Hayasaka, K., Himoro, M., Soto, W., Takada, G., Uyemura, K., Shimizu, N., Bird, T.D., Conneally, P.M., & Chance, P.F. (1993). Charcot-Marie-Tooth neuropathy type 1 is associated with mutations of the myelin Po gene. Nature Genetics 5, 31-34.

Hayasaka, K., Nanao, K., Tahara, M., Soto, W., Takada, G., Miura, M., and Uyemura, K. (1991). Isolation and sequence determination of cDNA coding the Major Structural Protein of Human Peripheral Myelin. Biochem. Biophy. Res. Commun. 180, 515-518.

Hirano, S., Kimoto, N., Shimoyama, Y., Hirohashi, S., & Takeichi, M. (1992). Identification of a neural α-Catenin as a key regulator of cadherin function and multicellular organization. Cell 70, 293-301.

Horlick, R.A., Stack, S.L., & Cooke, G.M. (1992). Cloning, expression, and tissue distribution of the gene encoding rat fibroblast growth factor receptor subtype 4. Gene 120, 291-295.

Huang, M., Lipfert, L., Cunningham, M., Brugge, J.S., Ginsberg, M.H., & Shattil, S.J. (1993). Adhesive ligand binding to integrin α11b stimulates tyrosine phosphorylation of novel protein substrates before phosphorylation of pp125 FAX. J. Cell Biol. 122, 473-483.

Hunkapiller, T., & Hood, L. (1989). Diversity of the immunoglobulin gene superfamily. Adv. Immunol. 44, 1-63.

Hynes, R.O. (1992). Integrins: Versatility, modulation and signaling in cell adhesion. Cell 69, 11-25.

Kadmon, G., Kowitz, A., Altevogt, P., & Schachner, M. (1990). Functional cooperation between the neural adhesion molecules L1 and N-CAM is carbohydrate dependent. J. Cell Biol. 110, 209-218.

Kirschner, A., & Ganser, A.L. (1980). Compact myelin exists in the absence of basic protein in the *shiverer* mutant mouse. Nature 283, 207-210.

Kulkens, T., Bolhius, P.A., Woltermann, R.A., Kemp, S. Nijenhuis, S., Valentijun, L.J., Hensels, G.W., Jennekens, F.G., Visser, M., Hoogendijk, J., & Baas, F. (1993). Deletion of the serine 34 codon from the major peripheral myelin protein Po gene in Charcot-Marie-Tooth disease type 1. Nature Genetics 5, 35-39.

Kuremund, V., Jungalawa, F.B., Fischer, G., et al. (1988). The L2/HNK-1 carbohydrate of neural cell adhesion molecules is involved in cell interactions. J. Cell Biol. 106, 213-223.

Lai, C., Brow, M.A., Nave, K.A., Noronha, A.B., Qualres, R.H., Bloom, F.E., Milner, R.J., & Sutcliffe, J.G. (1987). Two forms of 1B236/myelin-associated glycoprotein, a cell adhesion molecule for postnatal neural development, are produced by alternative splicing. Proc. Natl. Acad. Sci. USA 84, 3337-4341.

Lamperth, L., Menuelidis, L., & Webster, H.deF. (1991). Non myelin-forming perineuronal Schwann cells in rat trigeminal; ganglia express Po myelin glycoprotein mRNA during postnatal development. Mol. Res. 5, 177-181.

Lemke, G. (1988). Unwrapping the genes of myelin. Neuron 1, 535-543.

Lemke, G. & Axel, R. (1985). Isolation and sequence of a cDNA encoding the major structural protein of peripheral myelin. Cell 40, 501-508.

Lemke, G., Lamar, E., & Patterson, J. (1988). Isolation and analysis of the gene encoding peripheral myelin-protein zero. Neuron 1, 73-83.

Martini, R., Bollensen, E., & Schachner, M. (1988). Immunocytological localization of the major peripheral nervous system glycoprotein Po and L2/HNK-1 and L3 carbohydrate structures in developing and adult mouse sciatic nerve. Dev. Bio. 129, 330-338.

Martini, R., Mohajeri, M.H., Kasper, S., Giese, K.P., & Schachner, M. (1995a). Mice doubly deficient in the genes for Po and myelin basic protein show that both proteins contribute to the formation of the major dense line inperipheral nerve myelin. J. Neurosci. 15, 4488-4495.

Martini, R., Zielasek, J., Toyka, K.V., Giese, K.P., & Schachner, M. (1995b). Protein zero (PO)-deficient mice show myelin degeneration in peripheral nerves characteristic of inherited human neuropathies. Nature Genetics 11, 281-285.

Milner, P., Lovelidge, C.A., Taylor, W.A., & Hughes, R.A.C. (1987). Po myelin protein produces experimental allergic neuritis in Lewis rats. J. Neurol. Sci. 79, 275-285.

Mirsky, R., Winter, J., Abney, E., Pruso, R.M., Gavrilovic, J., & Roff, M.C. (1980). Myelin specific proteins and glycolipids in rat Schwann cells and oligodendrocytes in cultures. J. Cell Biol. 84, 483-494.

Nagafuchi, A., & Takeichi, M. (1988). Cell binding function of E-cadherin-associated protein: Similarity to vinculin and posttranscriptional regulation of expression. Cell 65, 849-857.

Nagafuchi, A., Takeichi, M., & Tsukita, S. (1991). The 102kd cadherin-associated protein: Similarity to vinculin and posttranscriptional regulation of expression. Cell 65, 849-857.

Newman, P.J., Berndt, M.C., Gorski, J., White, G.C., Lyman, S., Paddock, C., & Muller, W.A. (1990). PECAM-1 (CD31) cloning and relation to adhesion molecules of the immunoglobulin gene superfamily. Science 247, 1219-1222.

O'Shannessy, D.J., Willison, H.J., Inuzuka, T., Dobersen, M.J., & Aurles, R.H. (1985). The species distribution of nervous system antigens that react with anti-myelin-associated glycoprotein antibodies. J. Neuroimmunol. 9, 255-268.

Patel, P.I., & Lupski, J.R. (1994). Charcot-Marie-Tooth disease: A new paradigm for the mechanism of inherited disease. Trends in Genetics 10, 128-133.

Peters, A., Palay, S.L., & Webster, H. deF. (1991). *The Fine Structure of the Nervous System.* Oxford University Press, New York.

Pizzey, J., Rowett, L.H., Barton, C.H., Dickson, G., & Waslsh, F.S. (1989). Intercellular adhesion mediated by human muscle neural cell adhesion molecule: Effects of alternative exon usage. J. Cell Biol. 109, 3465-3476.

Poduslo, J.F. (1984). Regulation of Myelination: Biosynthesis of the major myelin glycoprotein by Schwann cells in the presence and absence of myelin assembly. J. Neurochem. 42, 493-503.

Recny, M.A., Luther, M.A., Knoppers, M.H., Neidhardt, E.A., Khandekar, S.S., Concino, M.F., Schimke, P.A., Francis, M.A., Moebius, U., Reinhold, B.B., Reinhold, U.N., & Reinharz, E.L. (1992). N-glycosylation is required for human CD2 immunoadhesion functions. J. Biol. Chem. 267, 22428-22434.

Roach, A., Boylan, K., Horvath, S., Prusiner, S.D., & Hood, L.E. (1983). Characterization of a cloned cDNA representing rat myelin basic protein: Absence of expression in *shiverer* mutant mice. Cell 34, 799-806.

Rutkowski, L., Needham, L., Frayer, K., Carson, D., McKhann, G., & Tennekoon, G.I. (1990). Evidence that secondary rat Schwann cells in culture maintain their differentiated phenotype. J. Neurochem. 54, 1895-1904.

Saavedra, R.A., Fors, L., Aebersold, R.H., Arden, B., Horvath, S., Sanders, J., & Hood, L. (1989). The myelin proteins of shark brain are similar to the myelin proteins of the mammalian peripheral nervous system. J. Mol. Evol. 29, 149-156.

Safranek, T.J., Lawrence, D.N., Kurland, L.T., Culver, D.H., Wiederholt, W.C., Hayner, N.S., Osterholm, M.T., O'Brien, P., & Hughes, J.M. (1991). Reassessment of the association between Guillain-BarrÇ syndrome and receipt of swine influenza vaccine in 1976-1977: Results of two-state study. Am. J. Epidemiol. 133, 940-952.

Sakamoto, Y., Kitamura, K., Yoshimura, K., Nishijima, T., & Uyemura, K. (1987). Complete amino acid sequence of Po-protein in bovine peripheral nerve myelin. J. Biol. Chem. 262, 4208-4214.

Salzer, J.L., & Colman, D.R. (1989). Mechanisms of cell adhesion in the nervous system: Role of the immunoglobulin gene superfamily. Dev. Neurosci. 11, 377-390.

Schneider-Schaulies, J., von Brunn, A., & Schachner, M. (1990). Recombinant peripheral myelin protein Po confers both adhesion and neurite outgrowth-promoting properties. J. Neurosci. Res. 27, 286-297.

Schonberger, L.B., Bregman, D.J., & Sullivan-Bolyai, J.Z. (1979). Guillain-BarrÇ syndrome following vaccination in the National Influenza Immunization Program, United States, 1976-1977. Am. J. Epidemio. 110, 105-123.

Sherman, L., Stocker, K.M., Morrison, R., & Ciment, G. (1993). Basic fibroblast growth factor (bFGF) acts intracellularly to cause the transdifferentiation of avian neural crest-derived Schwann cell precursors into melanocytes. Develop. 118, 1313-1326.

Suter, U., Welcher, A.A., Ôzcelik, T., Snipes, G.J., Kosaras, B., Francke, U., Billings-Gagliardi, S., Sidman, R.L., & Shooter, E.M. (1992a). Trembler mouse carries a point mutastion in a myelin gene. Nature 356, 241-244.

Suter, U., Welcher, A.A., & Snipes, G.J. (1992b). Progress in the molecular understandinng of hereditary peripheral neuropathies reveals new insights into the biology of the peripheral nervous system. TINS 16, 50-56.

Suzuki, M., Sakamoto, Y. Kitamura, K., Fukunaga, K., Yamamoto, H., Miyamoto, E., & Uyemura, K. (1990). Phosphorylation of Po glycoprotein in peripheral nerve myelin. J. Neurochem. 55, 1966-1971.

Thomas, F.P., Lebo, R.V., Ding, X., Latov, N., & Hays, A.P. (1993). Tomaculous neuropathy in Chromosome 1 Charcot-Marie-Tooth syndrome. Submitted.

Trapp, B.D., Kidd, G.J., Hauer, P., Mulrenin, E., Haney, C.A., & Andrews, S.B. (1995). Polarization of myelinating Schwann cell surface membranes: Role of Microtubules and Trans-Golgi Network. J. Neurosci. 15, 1797-1807.

Trapp, B.D., McIntyre, L.J., Quarles, R.H., Sternberger, N.H., & Webster, H.deeF. (1981). Immunocytochemical localization of Po-protein in Golgi membranes and myelin of developing rat Schwann cells. J. Cell Biol. 90, 1-6.

Uyemura, K., Kitamura, K., & Miura, M. (1992). Structure and Molecular Biology of Po Protein. In: *Myelin: Biology and Chemistry* (Martenson, R.E., Ed.). CRC Press, Boca Raton, FL, pp. 481-508.

Welcher, A.A., Suter, U., DeLeon, M. Snipes, G.J., & Shooter, E.M. (1991). A myelin protein is encoded by the homologue of a growth arrest-specific gene. Proc. Natl. Acad. Sci. USA 88, 7195-7199.

Wells, C.R., Saavedra, R., Inouye, H., & Kirschner, D.A. (1993). Myelin Po glycoprotein: Predicted structure and interactions of the extracellular domain. J. Neurochem. 61, 1987-1995.

Wiggins, R.C., & Morell, P. (1980). Phosphorylation and fucosylation of myelin protein *in vitro* by sciatic nerve from developing rats. J. Neurochem. 10, 627-634.

Williams, A.F., & Barclay, A.N. (1988). The immunoglobulin superfamily-domains for cell surface recognition. Ann. Rev. Immunol. 6, 381-405.

Wojciechowicz, D., Lu, C.F., Kurjan, J., & Lipke, P. (1993). Cell surface anchorage and ligand-binding domains of the *Saccharomyces cerevisiae* cell adhesion protein α-agglutinin, a member of the immunoglobulin superfamily. Mol. Cell. Biol. 13, 2554-2563.

Wong, M., & Filbin, M.T. (1994). The cytoplasmic domain of the myelin protein Po influences the adhesion of its extracellular domain. J. Cell Biol. 126, 367-370.

Wong, M.H., & Filbin, M.T. (1996). Dominant negative effect on adhesion by myelin Po protein truncated in its cytoplasmic domain. J. Cell Biol.

Yarden, Y., Escobedo, J.A., Kuang, W-J., Yang-Feng, T.L., Daniel, T.O., Tremble, P.M., Chen, E.Y., Ando, M.E., Harkins, R.N., Francke, U., Fried, V.A., Ullrich, A., & Williams, L.T. (1986). Structure of the receptor for platelet-derived growth factor helps define a family of closely related growth factor receptors. Nature 323, 226-232.

Yazaki, T., Miura, M., Asou, H., Kitamura, K., Toya, S., & Uyemura, K. (1992). Glycopeptide of Po protein inhibits homophilic cell adhesion: Competition assay with transformants and peptides. FEBS Lett. 307, 361-365.

Yazaki, T., Miura, M., Asou, H., Toya, S., & Uyemura, K. (1991). Myelin Po-protein expressed in C6 cells promotes neurite outgrowth. Biomed. Res. 12(4), 223-230.

Zachary, I., & Rosengunt, E. (1992). Focal adhesion kinase ($p125^{FAK}$). A point of convergence in the action of neuropeptides, integrins, and oncogenes. Cell 71, 891-894.

Zand, D., Hammer, J., Gould, R., & Quarles, R. (1991). High expression of the HNK-1/L2 carbohydrate epitope in the major glycoproteins of shark myelin. J. Neurochem. 57, 1076-1079.

Zhang, K., & Filbin, M.T. (1994). The role of the Ig-like domain disufide band of the myelin protein, Po, in its adhesion. J. Neurochem. 63, 367-370.

Zhang, K., Merazga, Y., & Filbin, M.T. (1996). Mapping the adhesive domains of the myelin protein Po. J. Neurosci. Res.

GPI-ANCHORED PROTEINS IN NEURAL CELL ADHESION

James L. Salzer, Charles L. Rosen, and Arie F. Struyk

Advances in Molecular and Cell Biology,
Volume 16, pages 193-222.

ISBN: 0-7623-0143-0.

I. INTRODUCTION

Cell interactions in the nervous system are mediated by a complex array of cell adhesion molecules (CAMs; Jessel, 1988; Edelman and Crossin, 1991; Reichardt and Tomaselli, 1991). Typically these CAMs are type one integral membrane proteins characterized by having an extracellularly disposed amino terminus, a single hydrophobic transmembrane segment, and an intracellular carboxy terminus of variable size. However, increasingly, neural adhesion molecules are being identified that do not traverse the bilayer but instead have a hydrophobic glycosylphosphatidyl-inositol (GPI) tail that anchors them to the external leaflet of the plasma membrane.

GPI-anchored proteins are expressed by organisms widely dispersed throughout phylogeny, being present in protozoans, yeast, slime mold, *Drosophila,* and man. GPI-anchored proteins also have a wide variety of functions serving as protozoal coat proteins, hydrolases, and growth factor and small molecule receptors (Low, 1989). However, while this glycolipid tail is not unique to neural cell adhesion molecules, it is striking that the most abundant GPI anchored molecules of the nervous system are cell adhesion molecules (Rosen et al., 1992) and are frequently members of the immunoglobulin gene superfamily (IgSF).

The reason why this mode of membrane anchoring is employed so often by neural adhesion molecules is not known but suggests an important role for the GPI anchor in the function of these proteins. In this chapter the structure and biosynthesis of the GPI anchor and its potential role in the function of cell adhesion molecules will be considered. Although the emphasis throughout will be on CAMs in the nervous system, illustrative examples of GPI anchored CAMs in other tissues, particularly in the immune system, will also be presented. Several excellent reviews on other aspects of GPI anchored proteins may be consulted for additional details not discussed here (Low, 1989; Cross, 1990; Ferguson, 1992; Englund, 1993).

II. STRUCTURE AND BIOSYNTHESIS OF THE GPI ANCHOR

The core structure of the GPI anchor was first elucidated for the membrane form of the variant surface glycoprotein of the African

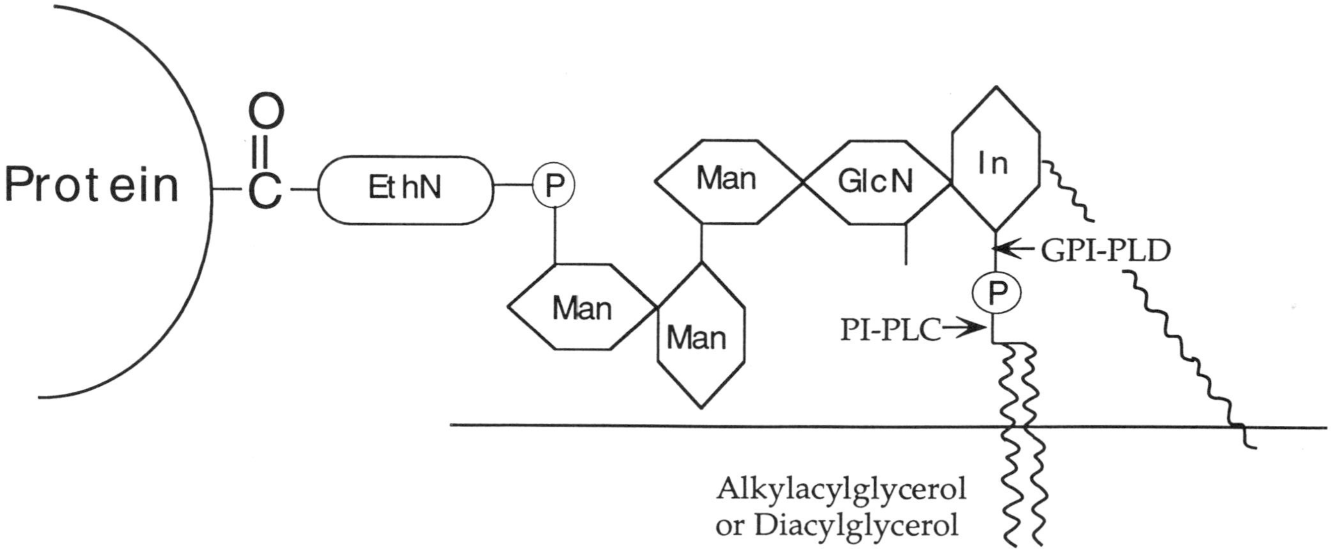

Figure 1. Structure of the GPI anchor. Proteins are linked to the GPI anchor at their carboxy termini. The anchor itself consists of a phosphoethanolamine attached to three mannoses, glucosamine, and inositol phosphate. The anchor inserts into the outer leaflet of the plasma membrane via an alkylacylglycerol or diacylglycerol moitey. Some GPI-anchored anchors are further modified by palmitylation at the 2- or 3-hydroxy residue of inositol. Also shown is the site of cleavage of PIPLC and GPI-PLD.

trypanosome (mfVSG). This protein is synthesized at enormous levels by trypanosomes (10^7 molecules/cell), which greatly facilitated its purification and characterization. It was subsequently found that the basic structure of the anchor, shown in Figure 1, is shared by all GPI-anchored proteins—indicative of remarkable conservation throughout evolution.

The GPI anchor is attached to the extreme carboxy terminus of proteins. The C-terminal amino acid of the protein is coupled via phosphoethanolamine to a linear tetrasaccharide comprised of three mannoses and one non acetylated glucosamine whose individual glycosidic linkages (6Manα 1-2Manα1-6Manα 1-4GlcNα1) are conserved in all GPI anchors. This glycan is O-glycosidically linked to the 6-hydroxyl group on the inositol ring of a diacyl or an alkylacyl glycerol phosphatidylinositol. Significant modifications of the core structure have been described. These include the presence of a palmitic acid on the inositol ring of human erythrocyte acetylcholinesterase (AChE) and decay accelerating factor (DAF), the presence of sialic acid on the anchor of the prion protein, and the presence of an extra phosphoethanolamine on the first mannose residue of Thy 1, AChE, and DAF (reviewed in Englund, 1993; Takeda and Kinoshita, 1995). The physiologic significance of these substitutions is not known. A practical consequence of the palmitylation of the insoitol ring, however, is that such modified anchors are resistant to cleavage by the bacterial enzyme phosphatidylinositol-specific phospholipase C (PIPLC) as described further ahead.

The anchor itself is built up sequentially by the addition of various sugars to phosphatidyl inositol, a process involving a series of enzymatic steps (Englund, 1993). Defects in as many as six different biosynthetic stages have been identified by genetic complementation studies using mutant cell lines that are deficient in the synthesis of all GPI-anchored proteins (Tartakoff and Singh, 1992). A defect in the biosynthesis of the GPI anchor at one of these steps is now appreciated to underlie the rare hematologic disorder, paroxysmal nocturnal hemoglobinuria (PNH). Patients with PNH are unable to synthesize GPI-anchored proteins on a proportion of their red blood cells that arise from a pluripotent stem cell carrying a somatic mutation (Rosse, 1990). The biosynthetic defect was recently found to result from a mutation of the enzyme that synthesizes N-acetylglucosaminyl phosphatidylinositol, the first intermediate in GPI anchor biosynthesis (Takeda et al., 1993). Among the missing

GPI-anchored proteins that are not synthesized as a result are decay accelerating factor and CD59, two proteins that normally protect red blood cells from the action of complement. Red blood cells from PNH patients are therefore particularly susceptible to autologous complement mediated hemolysis. It is not yet known whether comparable defects in the synthesis of GPI anchored proteins will account for clinical disorders of other tissues, including the nervous system.

Once the GPI anchor is synthesized, it is transferred en bloc to the carboxy terminus of the recipient protein (Englund, 1993). All GPI anchored proteins are initially synthesized with a C-terminal segment of predominately hydrophobic amino acids. Immediately after synthesis in the rough ER, this hydrophobic stretch of amino acids is cleaved off and the preformed glycolipid anchor is attached to the new carboxyl terminus. The addition of the anchor is believed to involve a nucleophilic attack catalyzed by a putative transamidase, in which the free amine group of ethanolamine displaces the carboxy terminal peptide of the protein (Englund, 1993). The signal that specifies GPI addition to a protein is comprised of two parts—a short (15-20 amino acids) hydrophobic COOH terminal peptide and a processing site containing two or three small amino acids that are located 10 to 12 amino acids amino terminal to the hydrophobic segment (Field and Menon, 1993; Kodukula et al., 1993).

III. CHARACTERIZATION OF THE MAJOR GPI-ANCHORED PROTEINS IN THE NERVOUS SYSTEM

A. Methods of Detection

Several methods have been used to detect GPI-anchored proteins. These include metabolic labeling of their anchors with radioactive ethanolamine or inositol. However, perhaps because of the relatively low abundance of these proteins and their long half lives (e.g., Lemansky et al., 1990; Moss and White, 1992), metabolic labeling of the GPI anchor of these proteins tends to be inefficient. Alternatively, the presence of a canonical hydrophobic segment at the extreme carboxy terminus of a deduced amino acid sequence has frequently been taken as presumptive evidence that a plasma membrane protein is GPI anchored.

The most commonly used criterion that a protein is GPI anchored is whether it can be released from the cell surface by PIPLC. This method is straightforward and generally reliable (Ikezawa, 1991), with the major limitation being that acylation of the anchor may render the protein insensitive to release. Moreover, PIPLC cleavage, when combined with Triton X-114 partitioning, is a very powerful and specific method for detecting and purifying GPI-anchored proteins (Lisanti et al., 1988). Triton X-114 is an anionic detergent that remains in solution at low temperature but separates into detergent and aqueous phases upon warming. These phases are enriched in hydrophobic (i.e., integral) and hydrophilic (i.e., peripheral) membrane proteins respectively (Bordier, 1981). GPI-anchored proteins normally partition into the detergent phase but with the loss of their lipid anchor after PIPLC cleavage, partition into and can be recovered from the aqueous phase instead (Lisanti et al., 1988).

B. The Most Abundant GPI-Anchored Proteins in the Nervous System are CAMs

We recently systematically characterized the major GPI-anchored proteins on purified populations of neurons (Rosen et al., 1992) and from defined regions of the rat brain (Struyk et al., 1992). To identify these proteins, we took advantage of: (1) their cell surface localization, (2) their sensitivity to cleavage with PIPLC, and (3) their conversion from a hydrophobic to a hydrophilic protein as a consequence of this cleavage. In these studies, primary cultures of neurons or brain membrane preparations were biotinylated, proteins were partitioned into the Triton X-114 detergent phase, the detergent phase was treated with PIPLC, and those proteins partitioning into the aqueous phase were precipitated, fractionated by SDS PAGE, and visualized with ^{125}I streptavidin after blotting onto nitrocellulose. This procedure results in a highly enriched preparation of GPI anchored proteins (Rosen et al., 1992). The results from such an analysis are shown in Figures 2 and 3.

To identify the various GPI-anchored proteins shown in Figures 2 and 3, we used immunoblotting and immunoprecipitation with antibodies to known GPI-linked proteins in the nervous system. In instances where we could not identify the protein with antibodies, we purified and obtained protein microsequence data for them

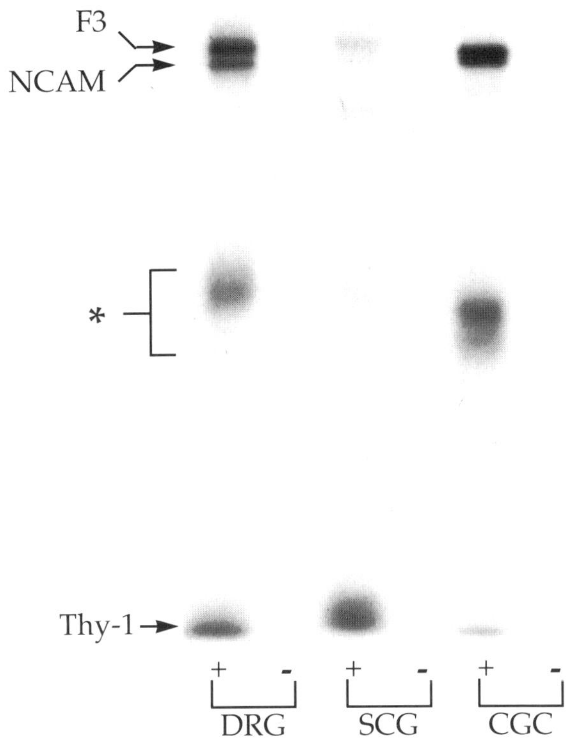

Figure 2. Differential expression of GPI-anchored proteins by primary neurons. Cultures of sensory neurons of dorsal root ganglia (DRG), sympathetic neurons isolated from superior cervical ganglia (SCG) and cerebellar granule cells (CGC) were biotinylated and solubilized with 1% Triton X-114. In each case, the detergent phase was then treated (+) or mock treated (−) with PIPLC for one hour at 37°C. Proteins released from the detergent phase into the aqueous phase were precipitated, separated by SDS PAGE, electroblotted, and probed with ^{125}I streptavidin. The major GPI-anchored proteins, corresponding to F3/F11, NCAM 120, a heterogeneous band of proteins (*) containing neurotrimin (and related proteins) and the CNTF receptor, and Thy 1 are indicated. The protein composition of each band was established by immunoprecipitation and immunoblotting with antibodies to cell adhesion molecules.

(Struyk et al., 1995). These studies led to the recognition that the major GPI-anchored neural proteins in the mammalian nervous system are adhesion molecules of the immunoglobulin gene superfamily, many, but not all of which had previously been identified. These include F3/F11 (Brümmendorf et al., 1989;

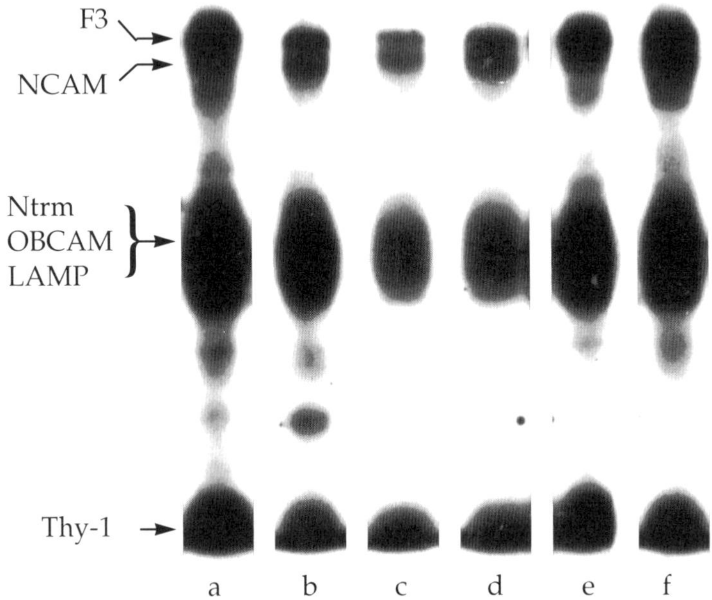

Figure 3. Expression of GPI-anchored proteins in different regions of the CNS. Membrane fractions from different regions of the adult rat CNS were prepared, biotinylated, extracted with 1% Triton X-114 and treated with PIPLC for one hour at 37°C. Proteins released from the detergent phase into the aqueous phase were precipitated, separated by SDS PAGE, electroblotted, and probed with ^{125}I streptavidin. The identity of the major GPI anchored proteins was established as above. Lanes correspond to cerebrum (a), cerebellum (b), brain stem (c), spinal cord (d), olfactory tract (e), and optic nerve (f). In addition to the major bands indicated, less abundant proteins with a Mr of approximately 35 kD and 45, whose identity has not been established, are visible in lanes a, b, e, and f.

Gennarini et al., 1989), which has a m wt of 135 kD, the GPI-anchored isoform of the neural cell adhesion molecule (N-CAM), which has a m wt of 120 kD (He et al., 1986), and Thy 1, which has a m wt of 25 kD (Tse et al., 1985).

The most abundant GPI-anchored proteins migrate as a broad band with a m wt of 65-70 kD. We have recently found that most of the proteins migrating in this broad band constitute a distinct subfamily of the IgSF whose members are comprised of three Ig domains. This subfamily includes the opiate binding cell adhesion molecule (OB-CAM; Schofield et al., 1989), a novel but highly related protein that we have termed neurotrimin (Struyk et al., 1995) and its chick homologue CEPU-1 (Spaltman and Brümmendorf, 1996), as well as the limbic system-associated protein (LAMP; Pimenta et al., 1995). Southern blotting studies suggest that this family contains several, as yet unidentified additional members (Struyk et al., 1995). Finally, the binding subunit of the ciliary neurotrophic factor receptor (CNTFR) is a 65 kD GPI anchored protein (Davis et al., 1991) that comprises a minor component of this band in the CNS (Rosen, Struyk, and Salzer, unpublished).

In addition, other less abundant GPI-anchored neural CAMs with a more restricted distribution or temporal pattern of expression have been described. These include TAG-1, a member of the IgSF that is transiently expressed on a subset of neurons (Furley et al., 1990), its chick homologue axonin-1 (Zuellig et al., 1992) and their structural homologues BIG-1 (Yoshihara et al., 1994) and BIG-2 (Yoshihara et al., 1995); several developmentally regulated proteoglycans (Herndon and Lander, 1990) including glypican (Karthikeyan et al., 1992) and cerebroglycan, a nervous system specific proteoglycan (Stipp et al., 1994); a transiently expressed 31 kD lectin on parallel fibers of the cerebellum (Kuchler et al., 1989); and two isoforms of T cadherin, a GPI-anchored member of the cadherin family which is expressed in neural and non-neural tissues (Ranscht and Dours-Zimmerman, 1991; Sacristán et al., 1993). Another GPI-anchored protein, the leucine rich oligodendrocyte myelin glycoprotein (OMgp), has been proposed, but not yet confirmed, to play a role in oligodendrocyte adhesion (Mikol et al., 1990).

A number of invertebrate neural CAMs that are GPI anchored include several members of the IgSF, namely the *Drosophila* or grasshopper proteins lacheson (Karlstrom et al., 1993), REGA-1 (Seaver et al., 1996) and amalgam (Harrelson and Goodman, 1988) and two isoforms of the *Aplysia* adhesion molecule apCAM (Mayford et al., 1992). Additional GPI-anchored proteins in *Drosophila* include the homophilic neural adhesion molecules chaoptin (Krantz and Zipursky, 1990) and fasciclin I (Hortsch and Goodman, 1990). The structures of those proteins whose sequences are known are shown diagramatically in Figure 4.

Several other GPI-anchored molecules in the nervous system are likely to have functions other than cell adhesion. These include 5' nucleotidase (Low, 1989), the prion protein (Stahl and Prusiner,

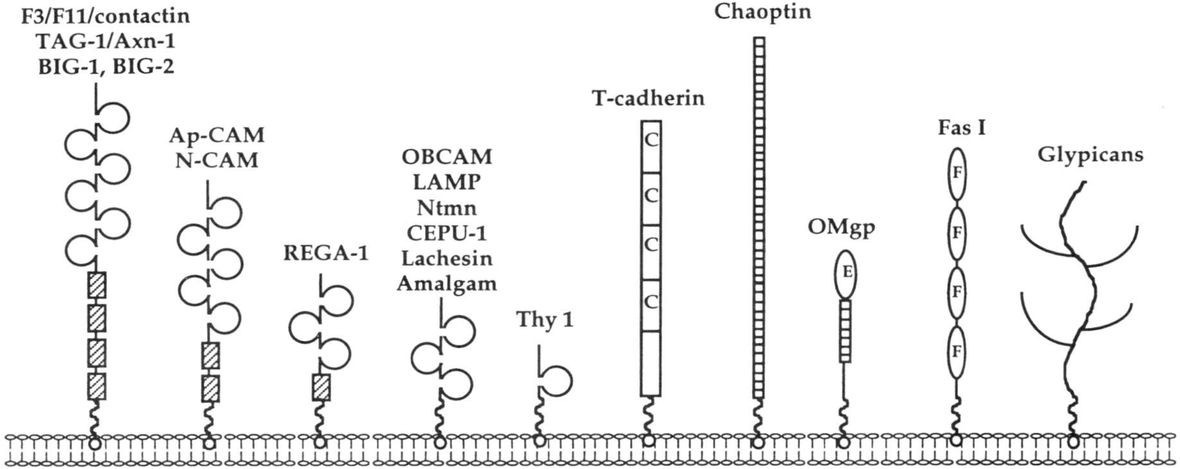

Figure 4. Schematic structures of GPI-anchored neural adhesion molecules. Many of the GPI-anchored proteins in the nervous system are Ig CAMs and contain one or more immunoglobulin homology units (shown as half circles) and fibronectin type III repeats (cross hatched boxes). Ig subfamilies consist of (1) six Ig domains and four fibronectin repeats, that is, F3/F11/contactin (Gennarini et al., 1989; Brümmendorf et al., 1989; Berglund and Ranscht, 1994), TAG-1/axonin-1 (Furley et al., 1990; Zuellig et al., 1992), BIG-1 (Yoshihara et al., 1994), and BIG-2 (Yoshihara et al., 1995); (2) five Ig domains and two fibronectin repeats, for example, Ap-CAM (Mayford et al., 1992) and N-CAM (Cunningham et al., 1987); (3) three Ig domains and one fibronectin repeat, for example, REGA-1 (Seaver et al., 1996); (4) three Ig domains, for example, OBCAM (Schofield et al., 1989), LAMP (Pimenta et al., 1995), neurotrimin (Struyk et al., 1995), CEPU-1 (Spaltmann and Brümmendorf, 1996), lachesin (Karlstrom et al., 1993), and amalgam (Seeger et al., 1988); and (5) one Ig domain, for example, Thy-1 (Williams and Gagnon, 1982). Other neuronal GPI-anchored proteins include: T-cadherin (Ranscht and Dours-Zimmerman, 1991) which contains four cadherin homology units (boxes with a C) and a fifth domain of partial homology; chaoptin (Krantz and Zipursky, 1990) and OMgp (Mikol et al., 1990) which contain multiple leucine rich repeats and, in the case of OMgp, a single EGF-like homology unit; fasciclin I (Hortsch and Goodman, 1990; Zinn et al., 1988) which contains four domains of internal homology that do not display significant homology to other proteins and the glypican family of proteoglycans which includes cerebroglycan, glypican, and several other proteins (see Lander et al., 1996 for a recent review).

1991), and binding subunit of the ciliary neurotrophic factor and glial-derived nerve trophic factor receptors (Davis et al., 1991; Treanor et al., 1996). In general, these proteins appear to be expressed at the cell surface at much reduced levels based on assays of the type shown in Figures 2 and 3. One potential limitation in determining their relative abundance by this method, however, is the possibility that the GPI anchors of these proteins could be differentially acylated and therefore differentially sensitive to cleavage by PIPLC.

IV. FUNCTIONAL STUDIES OF GPI-ANCHORED NEURAL ADHESION MOLECULES

GPI-anchored proteins in the nervous system frequently promote cell adhesion and are critical regulators of nerve fiber outgrowth during development. In one study (Chang et al., 1992), insect embryos were treated with PIPLC to cleave all the GPI-anchored proteins. As a consequence of this treatment, pioneer growth cones demonstrated developmentally specific errors in pathway choices. In this same study, fasciclin I expression levels were significantly disrupted, suggesting that it may have been an important target of the enzyme.

The role of GPI-anchored proteins in nerve fiber outgrowth has more frequently been investigated using purified proteins in *in vitro* model systems. For example, neuronal outgrowth has been measured on a substrate of purified protein or on transfected cells expressing a specific GPI-anchored protein at high levels. In studies of this type, NCAM 120 (Doherty et al., 1990), F3/F11 (Gennarini et al., 1991), LAMP (Pimenta et al., 1995), BIG-1 and BIG-2 (Yoshihara et al., 1994, 1995), and TAG-1 (Furley et al., 1990) were all found to promote nerve fiber outgrowth when presented as a substrate to appropriate neurons.

Frequently, this outgrowth promoting effect is mediated by heterophilic interactions (see chapter by J. Hemperley in this volume for additional details). Thus, axonin 1 binds to the Ig related transmembrane protein G4, itself a homologue of L1 (Kuhn et al., 1991). F3/F11, which promotes nerve fiber outgrowth, also appears to bind heterophilically (Faivre-Sarrailh et al., 1992) (although it may also mediate weak homophilical adhesion *in vitro;* Gennarini et al., 1991). Among its ligands are the related extracellular matrix proteins tenascin and restrictin, Ng-CAM, which may be the chick homologue of L1 (Brümmendorf and Rathjen, 1993) and the phosphatase RPTPB (Peters et al., 1995). TAG-1 also promotes binding via homophilic interactions and, when presented as a substrate, neurite outgrowth via heterophilic interactions with $\beta1$ integrins and an L1 like molecule (Felsenfeld et al., 1994).

Interestingly, many GPI-anchored proteins that promote nerve cell adhesion and neurite outgrowth, when presented to neurons as substrate molecules, mediate growth cone avoidance and inhibition of nerve fiber outgrowth when functioning as receptors at the surface of neurons encountering heterophilic ligands. For example, F3/F11 mediates the inhibition of nerve fibers encountering a restrictin substrate (Pesheva et al., 1993).

In fact, other studies suggest that the normal function of some GPI-anchored proteins may be to specifically *inhibit* nerve fiber outgrowth. For example, Thy 1 has been reported to reduce the extent of nerve fiber outgrowth on mature astrocytes (Tiveron et al., 1992). Another protein that functions as a "repulsion" molecule during development (by routing avian retinal nerve fibers to appropriate targets in the optic tectum; Bonhoeffer and Huf, 1982) is also GPI anchored (Walter et al., 1990).

Although all of the aforementioned proteins are members of the IgSF, at least one member of the cadherin family, T-cadherin, is also GPI anchored. Two isoforms of this protein have been identified; both are GPI anchored, differing only in a few amino acids at their extreme carboxy termini (Ranscht and Dours-Zimmerman, 1991; Sacristán et al., 1993). They are coexpressed in a limited number of tissue sites during development, including the embryonic nervous system (Sacristán et al., 1993). Like other cadherins, they mediate calcium dependent homophilic adhesion (Vestal and Ranscht, 1992). As other cadherins normally require an intact cytoplasmic domain to mediate effective adhesion (Nagafuchi and Takeichi, 1989; Ozawa et al., 1989), the mechanism by which T-cadherin promotes adhesion is not yet clear.

Taken together, these findings indicate that GPI-anchored proteins plays a critical role in mediating cell interactions in the nervous system.

V. POTENTIAL ROLE OF THE GPI ANCHOR FOR NEURAL CELL ADHESION MOLECULES

Because a functionally diverse group of proteins including a variety of cell surface hydrolases, membrane receptors, surface antigens on protozoans, and cell adhesion molecules, are GPI anchored, it is unlikely that this anchor serves a single function. In the nervous system, however, as was noted earlier, most GPI-anchored proteins appear to mediate cell-cell interactions (Rosen et al., 1992). This suggests that the GPI anchor may be particularly well suited to the function of adhesion molecules. Among the potential advantages of this anchor for CAMs are: (1) the ability to regulate the release (and thereby the abundance at the cell surface) of these molecules via anchor specific lipases, (2) regulation of the mobility of these molecules within the plane of the membrane, (3) use of the anchor to target these proteins to specific membrane domains, and (4) a role in cell signalling events. These potential functions of the anchor are considered in further detail below.

A. Release of GPI-anchored Proteins from the Cell Surface

Many GPI-anchored proteins exist as both integral membrane proteins and as soluble molecules. *In vivo,* Thy-1 (Almqvist and Carlsson, 1988), F3/F11, and NCAM 120 (but little of the transmembrane isoforms of NCAM; Durbec et al., 1992) have all been found in soluble form in the cerebrospinal fluid. *In vitro,* the GPI anchored form of NCAM, but not the transmembrane forms, was found to be released by an astroglial cell line in culture (He et al., 1987). Other GPI-anchored proteins that are also released by cultured cells into the media include TAG-1 (Furley et al., 1990) and its homologue axonin-1 (Stoeckli et al., 1989), and the prion protein (Borchelt et al., 1990).

These findings raise the possibility that these proteins are released from the cell surface as the result of cleavage of the glycolipid anchor by the activity of anchor specific phospholipases. In possible support of such a mechanism of release is the existence of two classes of anchor specific phospholipases: phosphatidylinositol-specific phospholipase C (PIPLC) and GPI-specific phospholipase D (GPI-PLD; see Figure 1 for their sites of action). Although cleavage of the anchor would be expected to result in the release of the protein from the cell membrane, as discussed ahead the actual physiologic role of these phospholipases is not clear at this time.

GPI-PLD activity is present at high levels in the serum. The source (or sources) of this GPI-PLD activity has not yet been established. Enzymatic activity has been detected in mast cells (Metz et al., 1991), in pancreatic islet cells (Metz et al., 1992), and in the brain (Hoener et al., 1990) and a cDNA for GPI-PLD has been isolated from the liver (Scallon et al., 1991). An immunohistochemical study also suggested that GPI-PLD may be expressed by specific neurons in the spinal cord and brain (Sesko and Low, 1991).

Although the presence of high levels of GPI-PLD in serum would at first appear to suggest a role for this enzyme in the release of GPI-anchored proteins, this remains as yet unproven. GPI-PLD is unable to cleave GPI-anchored proteins from the surface of intact cells; it is, however, effective on detergent treated cells (Low and Huang, 1991). Studies using a cDNA encoding GPI-PLD from the liver suggest that this enzyme may cleave the GPI anchor inside the cell rather than at the cell membrane (Scallon et al., 1991). Thus, transfection of cells with a cDNA encoding of GPI-PLD resulted in the release of alkaline phosphatase, a GPI anchored protein, into the media. Treatment of the cells with the recombinant enzyme released into the media, by contrast, had no effect on the release of alkaline phosphatase (Scallon et al., 1991). Of interest are recent findings of the presence of endogenous GPI-PLD mRNA and enzyme activity in cultured bone marrow cells and HeLa cells, and the demonstration that this activity results in the release of several GPI-anchored proteins (Brunner et al., 1994; Metz et al., 1994). It is not yet known whether this is an important mechanism of release in other cell types including neurons.

By contrast, PIPLC, in particular that purified from bacteria, is able to cleave most GPI-anchored proteins from the surface of intact cells. However, as noted previously, acylation of the inositol ring can render GPI anchored proteins resitant to cleavage by PIPLC (Toutant et al., 1990; Walter et al., 1990). This modification is generally thought to affect the GPI anchor of all proteins on the cell equivalently; its physiologic significance is not known. Much less is known about the PIPLC enzymes from eukaryotes. Enzymatic activity associated with mouse brain membranes (Fouchier et al., 1990) and at the surface of mouse 3T3 cells (Ting and Pagano, 1990) have been reported, but the enzymes involved have not been purified or cloned.

Although the existence of soluble forms might seem to suggest cleavage by a phospholipase, the actual mechanism by which GPI-anchored proteins are released from the cell surface has not been established. Other potential mechanisms for their release from the cell surface include proteolytic cleavage at or near the plasma membrane or the release by cells of small membrane vesicles containing proteins that are still GPI anchored (Ferguson, 1992). In addition, soluble forms of these proteins that are found in various tissue fluids such as the CSF could reflect the secretion of alternatively spliced isoforms for these proteins, as found for NCAM (Gower et al., 1988) and suggested for axonin-1 (Stoeckli et al., 1991). In the case of NCAM 120, the protein released by cultured cells does contain ethanolamine, suggesting that it is indeed released by a phospholipase rather than by a protease (He et al., 1987). Clearly, additional studies will be needed to elucidate what may be a variety of mechanisms for converting these membrane proteins into soluble proteins.

Whatever mechanism(s) are involved, release of these proteins from the cell surface has several potentially important functional consequences. Cleavage of the GPI anchor, if under physiologic control, may provide a mechanism for regulating the abundance of these proteins at the cell surface. This, in turn, could determine the adhesive properties of the cell surface and possibly regulate a transition between adherence and non-adherence. Evidence for the regulated release of GPI-anchored proteins has come from several studies. For example, the relative abundance of the soluble form of the *Drosophila* protein fasciclin I is modulated during embryogenesis, suggesting that cleavage of its GPI anchor may be developmentally regulated (Hortsch and Goodman, 1990). In other studies, activation of a variety of cells by different growth factors resulted in the rapid down regulation at the cell surface (Ishihara et al., 1987; Lisanti et al., 1986; Kubota et al., 1990) and associated release into the media as soluble proteins (Chan et al., 1988; Roberts et al., 1990) of a number of different GPI-anchored proteins. Such a rapid down modulation could reflect changes in the activity of an endogenous phospholipase(s).

In addition, the released soluble protein may itself be biologically active. Soluble forms of GPI-anchored proteins could bind to their cognate receptor, compete for binding sites, and have anti-adhesive effects (He et al., 1987; Furley et al., 1990). Evidence for such an anti-adhesive effect was found in studies in which high concentrations of axonin-I inhibited neurite fasciculation (Stoeckli et al., 1991). Also, the soluble form of the protein might promote neurite outgrowth directly, nonmechanically, by activating the cell. Striking evidence of such an effect was provided by a study in which soluble F3, and cerebrospinal fluid containing physiologic concentrations of soluble F3, was found to enhance neurite outgrowth directly (Durbec et al., 1992). These results suggest that GPI-anchored proteins have the potential to be released locally as soluble molecules, but to function throughout the entire nervous system as critical determinants of axon outgrowth.

Finally, a recent study (Barbon et al., 1995) suggests that the presence of the GPI anchor may have a significant influence on the conformation of the extracellular domain of the protein and may thereby regulate its functions.

B. Planar Mobility of GPI-anchored Proteins

Photobleaching studies have shown that GPI-linked proteins are frequently more mobile in the plane of the membrane than transmembrane proteins and have lateral diffusion coefficients approaching those of glycolipids (Ishihara et al., 1987; Noda et al., 1987). It was initially assumed that GPI-anchored proteins are more mobile within the plane of the membrane because they are only anchored to the outer leaflet of the plasma membrane and are therefore not constrained by interactions with the cytoskeleton. The actual mechanisms that account for the increased planar mobility of GPI-anchored proteins appear to be more complex however. Jacobson and his colleagues have presented evidence that the GPI linkage contributes only marginally to the higher mobility of GPI-anchored proteins and that the major determinant of the high mobility of these proteins is the lack of interaction of the extracellular domain of these proteins with other cell surface components (Zhang et al., 1991).

Photobleaching studies of GPI-anchored cell adhesion molecules at the surface of neurons have not yet been done and it is therefore not known whether the increase in planar mobility observed for some GPI-anchored molecules will also be true for neural CAMs such as F3/F11 and TAG-1. Provided that this turns out to be the case, increased lateral diffusion could promote stable, but fluid adhesion between the plasma membranes of cells in contact. This might allow closely apposed plasma membranes to more easily slip past one another while still maintaining contact. For example, as nerve fibers elongate through a fascicle of other fibers during development or as the myelin sheath rapidly spirals around an axon, membranes of different cells are required to rapidly extend against each other.

The GPI anchor does permit proteins to widely diffuse throughout the plane of the membrane, in contrast to transmembrane proteins which are confined to membrane microdomains presumably via their cytoskeletal interactions (Edidin and Stroynowski, 1991). Together with their increased lateral diffusion, this property of GPI-anchored proteins appears to promote adhesive interactions by facilitating the rapid recruitment of these proteins, at high concentration, into sites of cell contact. As the cell adhesion mediated by various members of the IgSF, such as NCAM, is particularly concentration dependent (Hoffman and Edelman, 1983; Doherty et al., 1990), increases in local concentration of adhesion molecules could have dramatic effects on cell adhesion. Consistent with this notion, binding of lymphocytes to an artificial monolayer containing low concentrations of a counter receptor (LFA-3) was more effective to the GPI-anchored isoform of this receptor (which was highly mobile under the experimental conditions) than to the transmembrane isoform (which was essentially immobile; Chan et al., 1991). These authors argue that increased lateral diffusion of cell surface receptors enhances the force of cell adhesion by promoting accumulation of ligands in the cell contact area and by increasing the rate of receptor-ligand bond formation (Chan et al., 1991).

C. The GPI Anchor as a Membrane Targeting Signal

The GPI linkage has been proposed to be a signal for sorting of proteins to the apical surface of polarized epithelial cells (Lisanti et al., 1989a). The apical and basal lateral surfaces of polarized epithelial cells, such as the Madin-Darby canine kidney (MDCK) cells, contain distinct sets of plasma membrane proteins (Simons and Wandinger-Ness, 1990). Characterization of the distribution of endogenous GPI-anchored proteins in MDCK cells revealed that they are preferentially localized to the apical surface of these cells (Lisanti et al., 1988). These results suggested that the GPI anchor is an apical targeting signal. Consistent with this hypothesis, heterologous GPI-anchored proteins, but not their transmembrane counterparts, were preferentially distributed to the apical membrane after transfection of MDCK cells (Brown et al., 1989). For example, the GPI-anchored form of NCAM was expressed at the apical surface of transfected MDCK cells, whereas the transmembrane forms of NCAM were expressed basolaterally (Powell et al., 1991).

It has been suggested that the GPI anchor could have a similar targeting function in the nervous system (Lisanti and Rodriguez-

Boulan, 1990). Axons and dendrites are specialized for distinct functions and some proteins are known to be differently expressed in these processes (Prochiantz, 1995). Indeed, in some studies, GPI-linked proteins have been reported to be enriched on axons, which resembles the apical domain of epithelial cells in some aspects (Dotti and Simons, 1990). Thy-1 and TAG-1, for example, have been reported to be present on axons, while being excluded from cell bodies and dendrites (Dodd et al., 1988; Dotti et al., 1991). However, other studies suggest that the distribution of Thy-1 and other GPI-anchored proteins *in vivo* may be quite complex and significantly modified as the result of specific cell-cell interactions (Xue et al., 1991; Faivre-Sarrailh et al., 1992). Future studies will be needed to clarify whether this is an important role for the GPI anchor *in vivo*.

The mechanism(s) accounting for the specific targeting of GPI-anchored proteins to the apical surface of polarized cells has not yet been elucidated. Sorting of apical and basolateral membrane components is likely to be accomplished by packaging proteins into distinct carrier vesicles that form in the trans-Golgi network (Simons and Wandinger-Ness, 1990). Apical membranes are also enriched in glycosphingolipids and it has been proposed that they associate in the Golgi and are sorted coordinately with apically targetted proteins such as GPI-anchored proteins (Simons and Wandinger-Ness, 1990). In support of such an association, GPI-anchored proteins, together with sphingolipids but not basolateral proteins, form an insoluble aggregate in the Golgi apparatus (Brown and Rose, 1992; Moss and White, 1992). The signal that subsequently directs the sorting of such aggregates is not yet known.

D. Clustering of GPI-anchored Proteins

Following antibody staining, the plane of the membrane in GPI-anchored proteins do not have a uniform distribution at the cell surface, but rather are clustered in microinvaginations of the plasma membrane referred to as caveolae. Caveolae appear as invaginated pits on the surface of a variety of non-neuronal cells with a diameter of about 50 nm and are believed to concentrate and mediate the uptake of small molecular weight proteins by a non-clathrin coated pit mechanism (Anderson, 1993). One example is the receptor for folate, which is GPI anchored, and mediates the sequestration and uptake of this vitamin. Caveolae are capable of transiently closing and may then be acidified. In response to low pH, the folate is believed to dissociate and be transported into the cytoplasm (Anderson et al., 1992). Uptake by such a mechanism has been referred to as potocytosis to distinguish it from clathrin mediated endocytosis (Anderson et al., 1992).

It was originally believed that GPI-anchored proteins in non-neuronal cells are concentrated in caveolae. More recent studies suggest that these proteins are not actually concentrated in caveolae but rather may be present in an annulus around the caveolae (Schnitzer et al., 1995) or, alternatively, have a uniform distribution at the cell surface and become associated with caveolae only secondarily, as a consequence of antibody cross-linking (Mayor et al., 1995). As neurons do not have morphologically identifiable caveolae or express caveolin (e.g., Gorodinsky and Harris, 1995), neuronal GPI-anchored proteins cannot be associated with this structure. Rather, they may be enriched, as in other cell types, in detergent insoluble rafts of glycolipids (Gorodinsky and Harris, 1995; Olive et al., 1995). Other recent data, however, suggest that this enrichment may be artefactual and results from detergent extraction (Mayor and Maxfield, 1995). Thus, it is presently uncertain whether GPI-anchored proteins are actually concentrated in a specialized membrane domain in neurons, or in any other cell type, and whether they are localized in proximity to and redistribute into, the caveolae of non-neuronal cells (see Parton and Simons, 1995 for a brief review). Further studies will be needed to clarify these important questions.

Although their precise distribution remains controversial, GPI-anchored proteins appear to be excluded from clathrin-coated pits. Consequently, they may reside at the cell surface longer and turn over more slowly than do transmembrane proteins. Thy-1 and F3/F11, for example, appear to have very long half-lives at the cell surface (Lemansky et al., 1990; Moss and White, 1992). A GPI-anchored form of CD4 was also found to turn over three times more slowly than its transmembrane counterpart (Keller et al., 1992). However, as previously discussed, the activity of endogenous phospholipases and proteases may be the critical determinants of their longevity at the cell surface.

E. Potential Role of GPI-anchored Proteins in Cell Signaling Events

Several lines of evidence have indirectly implicated GPI-anchored proteins in signal transduction. This was originally suggested by studies in which cross linking of various GPI-anchored proteins of T lymphocytes with antibodies (Kroczek et al., 1986; Seaman et al., 1991; Su et al., 1991) or lectins (Presky et al., 1990) resulted in T cell activation. This activation depends on the presence of a GPI anchor on the cross linked protein (Robinson, 1991; Su et al., 1991).

Potentially, this activation could reflect interactions of these GPI anchored proteins with signal transducing proteins of the T cell receptor complex (Robinson, 1991). Indeed, recent reports suggest an interaction of GPI anchored proteins with various cytoplasmic tyrosine kinases. Initial evidence for such an interaction was provided by genetic analysis of mutations of the *Drosophila* genes for fasciclin I and cAbl (a cytoplasmic tyrosine kinase). Null mutations for either of these genes have no obvious phenotype. However, striking abnormalities of growth cone guidance were observed in the corresponding double mutant, suggesting a possible functional interaction between these two proteins (Elkins et al., 1990). More direct evidence for such an interaction has emerged from studies in which cross linking of various GPI-anchored proteins of lymphocytes was shown to activate tyrosine phosphorylation (Hsi et al., 1989; Stefanová et al., 1991). Several groups have reported that a variety of cytoplasmic tyrosine kinases can be coprecipitated with GPI-anchored proteins (Stefanová et al., 1991; Thomas and Samelson, 1992) or are coenriched with them in non-ionic detergent insoluble complexes (Dráberová and Dráber, 1993). This association likewise requires the presence of an intact GPI anchor (Stefanová et al., 1991; Thomas and Samelson, 1992). Consistent with these findings, recent studies of the neuronal GPI-anchored protein F3/F11/contactin have demonstrated its association with the cytoplasmic tyrosine kinase fyn after antibody cross-linking (Olive et al., 1995; Zisch et al., 1995).

The mechanism for these reported interactions of cytoplasmic kinases with GPI anchored proteins is, as yet, unclear. Since the GPI anchor is present only in the external leaflet of the plasma membrane and cytoplasmic kinases associate with the inner leaflet, it must be assumed that one or more intermediary proteins are involved. A similar mechanism for activation of a cytoplasmic kinase by an intermediary protein was recently described in the case of CNTF (Stahl et al., 1994) and GDNF (Treanor et al., 1996). In the case of neural adhesion molecules, potential transmembrane proteins that might mediate such an effect include L1 (Olive et al., 1995) and a recently described 140 kD protein (Peles et al., 1995).

A second mechanism by which GPI-anchored proteins might activate cells is through the generation of second messengers via the cleavage of the anchor by endogenous phospholipases. Cleavage by PIPLC generates diacylglycerol, which could potentially activate protein kinase C associated with the cytoplasmic leaflet. Cleavage by GPI-PLD generates phosphatidic acid (and inositol glycan) which have been reported to have Ca^{++} mobilizing, mitogenic, and insulin mimetic activities (briefly reviewed in Brunner et al., 1994). At the present time, however, a role for such metabolites of the GPI anchor in cell signaling is only conjectural.

VI. CONCLUSIONS AND PERSPECTIVES

A growing number of cell adhesion molecules in the nervous system are GPI-anchored, suggesting an important role for this structure in the function of these proteins. Of particular interest is the ability of GPI-anchored neural adhesion molecules to regulate the direction and extent of nerve fiber outgrowth. While significant progress has been made in demonstrating their role in promoting cell adhesion and identifying their counter receptors in recent years, important questions remain. In particular, the role of these CAMs in cell signaling and the role of the GPI anchor in regulating their targeting, surface abundance, conformation, and planar mobility are not well understood.

An emerging and potentially unifying theme is the accumulation of many GPI-anchored proteins into glycolipid rich regions of the plasma membrane and their association with subjacent signaling molecules following antibody cross-linking. The extent to which ligand interactions similarly regulate the distribution and signaling function of these proteins is not yet known. Future studies should clarify the mechanisms by which GPI-anchored cell adhesion molecules regulate neurite outgrowth and the precise signaling pathways that are involved.

ACKNOWLEDGMENTS

We thank Martin Low for his generous gift of PIPLC during the course of these studies. Work by the authors cited in this chapter was supported by a grant from the American Paralysis Association and NIH grant NS

26001. C. Rosen and A. Struyk were supported by Medical Scientist training grant 5T32 GM07308 from NIGMS. J. Salzer is a recipient of an Irma T. Hirschl Career Scientist Award.

REFERENCES

Almqvist, P. & Carlsson, S.R. (1988). Characterization of a hydrophilic form of Thy-1 purified from human cerebrospinal fluid. J Biol Chem. 263, 12709-12715.

Anderson, R.G.W. (1993). Plasmalemmal caveolae and GPI-anchored membrane proteins. Curr. Opin. Cell Biol. 5, 647-652.

Anderson, R.G.W., Kamen, B.A., Rothberg, K.G., & Lacey, S.W. (1992). Potocytosisis: Sequestration and transport of small molecules by caveolae. Science. 255, 410-411.

Barboni, E., Rivero, B.P., George, A.J., Martin, S.R., Renoup, D.V., Hounsell, E.F., Barber, P.C., & Morris, R.J. (1995). The glycophosphatidylinositol anchor affects the conformation of Thy-1 protein. J. Cell Sci. 108, 487-497.

Berglund, E.O., & Ranscht, B. (1994). Molecular cloning and *in situ* localization of the human contactin gene (CNTN1) on chromosome 12q11-q12. Genomics 21, 571-582.

Bonhoeffer, F. & Huf, J. (1982). In vitro experiments on axon guidance demonstrating an anterior-posterior gradient on the tectum. Embo J. 1, 427-431.

Borchelt, D.R., Scott, M., Taraboulos, A., Stahl, N., & Prusiner, S.B. (1990). Scrapie and cellular prion proteins differ in their kinetics of synthesis and topology in cultured cells. J Cell Biol. 110, 743-752.

Bordier, C. (1981). Phase separation of integral membrane proteins in Triton X-14. J. Biol. Chem. 256, 1604-1607.

Brown, D.A., Crise, B., & Rose, J.K. (1989). Mechanism of membrane anchoring affects polarized expression of two proteins in MDCK cells. Cell 62, 1499-1501.

Brown, D.A. & Rose, J.K. (1992). Sorting of GPI-anchored proteins to glycolipid-enriched membrane subdomains during transport to the apical cell surface. Cell 68, 533-544.

Brümmendorf, T. & Rathjen, F.G. (1993). Axonal glycoproteins with immunoglobulin- and fibronectin type III-related domains in vertebrates: Structural features, binding activities, and signal transduction. J. Neurochem. 61, 1207-1219.

Brümmendorf, T., Wolff, J.M., Frank, R., & Rathjen, F. (1989). Neural celll recognition molecule F11: Homology with Fibronectin type III and immunoglobulin type C domains. Neuron 2, 1351-1361.

Brunner, G., Metz, C.N., Nguyen, H., Gabrilove, J., Patel, S.R., Davitz, M.A., Rifkin, D.B., & Wilson, E.L. (1994). An endogenous glycosylphosphatidylinositol-specific phospholipase D releases basic fibroblast growth factor-heparan sulfate proteoglycan complexes from human bone marrow cultures. Blood.

Chan, B.L., Lisanti, M.P., Rodriguez, B.E., & Saltiel, A.R. (1988). Insulin-stimulated release of lipoprotein lipase by metabolism of its phosphatidylinositol anchor. Science 241, 1670-1672.

Chan, P.Y., Lawrence, M.B., Dustin, M.L., Ferguson, L.M., Golan, D.E., & Springer, T.A. (1991). Influence of receptor lateral mobility on adhesion strengthening between membranes containing LFA-3 and CD2. J Cell Biol. 115, 245-255.

Chang, W.S., Serikawa, K., Allen, K., & Bentley, D. (1992). Disruption of pioneer growth cone guidance in vivo by removal of glycosyl-phosphatidylinositol-anchored cell surface proteins. Devel. 114, 507-519.

Cross, G.A.M. (1990). Glycolipid anchoring of plasma membrane proteins. Ann. Rev. Cell Biol. 6, 1-39.

Cunningham, B.A., Hemperly, J.J., Murray, B.A., Prediger, E.A., Brackenbury, R., & Edelman, G. (1987). Neural cell adhesion molecule: Structure, immunoglobulin-like domains, cell surface modulation, and alternative splicing. Science 236, 799-806.

Davis, S., Aldrich, T.H., Valenzuela, D.M., Wong, V.V., Furth, M.E., Squinto, S.P., & Yancopoulos, G.D. (1991). The receptor for ciliary neurotrophic factor. Science 253, 59-63.

Dodd, J., Morton, S.B., Karagogeos, D., Yamamoto, M., & Jessell, T.M. (1988). Spatial regulation of axonal glycoprotein expression on subsets of embryonic spinal neurons. Neuron 1, 105-116.

Doherty, P., Fruns, M., Seaton, P., Dickson, G., Barton, C.H., Sears, T.A., & Walsh, F.S. (1990). A threshold effect of the major isoforms of NCAM on neurite outgrowth. Nature 343, 464-466.

Dotti, C.G., Parton, R.G., & Simons, K. (1991). Polarized sorting of glypiated proteins in hippocampal neurons. Nature 349, 158-161.

Dotti, C.G. & Simons, K. (1990). Polarized sorting of viral glycoproteins to the axon and dendrites of hippocampal neurons in Culture. Cell 62, 63-72.

Dráberová, L. & Dráber, P. (1993). Thy-1 glycoportein and src-like protein-tyrosine kinase $p53/p56^{lyn}$ are associated in large detergent-resistant complexes in rat basophilic leukemia cells. Proc. Natl. Acad. Sci. USA 90, 3611-3615.

Durbec, P., Gennarini, G., Goridis, C., & Rougon, G. (1992). A soluble form of the F3 neuronal cell adhesion molecule promotes neurite outgrowth. J Cell Biol. 117, 877-888.

Edelman, G.M. & Crossin, K.L. (1991). Cell adhesion molecules: Implications for a molecular histology. Ann Rev Biochem. 60, 155-90.

Edidin, M. & Stroynowski, I. (1991). Differences between the lateral organization of conventional and inositol phospholipid-anchored membrane proteins: A further definition of micrometer scale membrane domains. J Cell Biol. 112, 1143-1150.

Elkins, T., Zinn, K., McAllister, L., Hoffmann, F.M., & Goodman, C.S. (1990). Genetic analysis of a Drosophila neural cell adhesion molecule: Interaction of fasciclin I and Abelson tyrosine kinase mutations. Cell 60, 565-575.

Englund, P.T. (1993). The structure and biosynthesis of glycosyl phosphatidylinositol protein anchors. Annu. Rev. Biochem. 62, 121-138.

Faivre-Sarrailh, C., Gennarini, G., Goridis, C., & Rougon, G. (1992). F3/F11 cell surface molecule expression in the developing mouse cerebellum is polarized at synaptic sites and within granule cells. J. Neurosci. 12, 257-267.

Felsenfeld, D.P., Hynes, M.A., Skoler, K.M., Furley, A.J., & Jessel, T.M. (1994). TAG-1 can mediate homophilic binding, but neurite outgrowth on TAG-1 requires an L1-like molecule and b1 integrins. Neuron 12, 675-90.

Ferguson, M.A.J. (1992). Glycosyl-phosphatidylinositol membrane anchors: The tale of a tail. Biochem. Soc. Trans. 20, 243-256.

Field, M.C. & Menon, A.K. (1993). Glycolipid anchoring of cell surface proteins. In: *Lipid Modifications of Proteins* (Schlesinger, M.J., Ed.). Boca Ratan, FL, CRC Press, pp. 84-134.

Fouchier, F., Baltz, T. & Rougon, G. (1990). Identification of glycosylphosphatidylinositol-specific-phospholipase C in mouse brain membranes. Biochem. J. 269, 321-327.

Friedman, G.C., & O'Leary, D.D.M. (1996). Eph receptor tryosine kinases and their ligands in neural development. Curr. Opin. Neurobiol. 6, 127-133.

Furley, A.J., Morton, S.B., Manalo, D., Karagogeos, D., Dodd, J., & Jessell, T.M. (1990). The axonal glycoprotein TAG-1 is an immunoglobulin superfamily member with neurite outgrowth-promoting activity. Cell 61, 157-170.

Gennarini, G., Cibelli, G., Rougon, G., Mattei, M.-G., & Goridis, C. (1989). The mouse neuronal cell surface protein F3: A phosphatidylinositol-anchored member of the immunoglobulin superfamily related to chicken contactin. J. Cell Biol. 109, 775-778.

Gennarini, G., Durbec, P., Boned, A., Rougon, G., & Goridis, C. (1991). Transfected F3/F11 neuronal cell surface protein mediates intercellular adhesion and promotes neurite outgrowth. Neuron 6, 595-606.

Gorodinsky, A., & Harris, D.A. (1995). Glycolipid-anchored proteins in neuroblastoma cells form detergent-resistant complexes without caveolin. J. Cell Biol. 129, 619-627.

Gower, H.J., Barton, C.H., Elsom, V.L., Thompson, J., Moore, S.E., Dickson, G., & Walsh, F.S. (1988). Alternative splicing generates a secreted form of N-CAM in muscle and brain. Cell 55, 955-964.

Harrelson, A.L. & Goodman, C.S. (1988). Growth cone guidance in insects: Fasciclin II is a member of the immunoglobulin superfamily. Science 242, 700-708.

He, H.T., Barbet, J., Chaix, J.C., & Goridis, C. (1986). Phosphatidylinositol is involved in the membrane attachment of NCAM-120, the smallest component of the neural cell adhesion molecule. Embo J. 5, 2489-2494.

He, H.T., Finne, J., & Goridis, C. (1987). Biosynthesis, membrane association, and release of N-CAM-120, a phosphatidylinositol-linked form of the neural cell adhesion molecule. J. Cell Biol. 105, 2489-2500.

Herndon, M.E. & Lander, A.D. (1990). A diverse set of developmentally regulated proteoglycans is expressed in the rat central nervous system. Neuron 4, 949-961.

Hoener, M.C., Stieger, S., & Brodeck, U. (1990). Isolation and charcterization of a phosphatidylinositol-glycan-specific phospholipase D from bovine brain. Eur. J. Biochem. 190, 593-601.

Hoffman, S. & Edelman, G.M. (1983). Kinetics of homophilic binding by E and A forms of the neural cell adhesion molecule. Proc. Natl. Acad. Sci. USA 80, 5762-5766.

Hortsch, M. & Goodman, C.S. (1990). Drosophila fasciclin I, a neural cell adhesion molecule, has a phosphatidylinositol lipid membrane anchor that is developmentally regulated. J. Biol. Chem. 265, 15104-9.

Hsi, E.D., Siegel, J.N., Minami, Y., Luong, E.T., Klausner, R.D., & Samelson, L.E. (1989). T cell activation induces rapid tyrosine phosphorylation of a limited number of cellular substrates. J. Biol. Chem. 264, 10836-42.

Ikezawa, H. (1991). Bacterial PIPLCs—unique properties and usefulness in studies on GPI anchors. Cell Biol. Int. Rep. 15, 1115-1131.

Ishihara, A., Hou, Y., & Jacobson, K. (1987). The Thy-1 antigen exhibits rapid lateral diffusion in the plasma membrane of rodent lymphoid cells and fibroblasts. Proc. Natl. Acad. Sci. USA 84, 1290-1293.

Ishihara, M., Fedarko, N.S., & Conrad, H.E. (1987). Involvement of phosphatidylinositol and insulin in the coordinate regulation of proteoheparan sulfate metabolism and hepatocyte growth. J. Biol. Chem. 262, 4708.

Jessel, T.M. (1988). Adhesion molecules and the hierarchy of neural development. Neuron 1, 3-13.

Karlstrom, R.O., Wilder, L.P., & Bastiani, M.J. (1993). Lachesin: An immunoglobulin superfamily protein whose expression correlates with neurogenesis in grasshopper embryos. Development 118, 509-522.

Karthikeyan, L., Maurel, P., Rauch, U., Margolis, R.K., & Margolis, R.U. (1992). Cloning of a major heparan sulfate proteoglycan from brain and identification as the rat form of glypican. Biochem. Biophys. Res. Commun. 188, 395-401.

Keller, G.-A., Siegel, M.W., & Caras, I.W. (1992). Endocytosis of glycophospholipid-anchored and transmembrane forms of CD4 by different endocytic pathways. Embo J. 11, 863-874.

Kodukula, K., Gerber, L.D., Amthauer, R., Brink, L., & Udenfriend, S. (1993). Biosynthesis of glycosylphosphatidylinositol (GPI)-anchored membrane proteins in intact cells: Specific amino acid requirements adjacent to the site of cleavage and GPI attachment. J. Cell Biol. 120, 657-664.

Krantz, D.E. & Zipursky, S.L. (1990). *Drosophila chaoptin,* a member of the leucine-rich repeat family, is a photoreceptor cell-specific adhesion molecule. EMBO J. 9, 1969-19977.

Kroczek, R.A., Gunter, K.C., Germain, R.N., & Shevach, E.M. (1986). Thy-1 functions as a signal transduction molecule in T lymphocytes and transfected B lymphocytes. Nature 322, 181-184.

Kubota, H., Okazaki, H., Onuma, M., Kano, S., Hattori, M., & Minato, N. (1990). Identification and gene cloning of a new phosphatidylinositol-linked antigen expressed on mature lymphocytes. Down-regulation by lymphocyte activation. J Immunol. 145, 3924-3931.

Kuchler, S., Rougon, G., Marschal, P., Lehman, S., Reeber, A., Vincendon, & Zanetta, J.-P. (1989). Location of a transiently expressed glycoprotein in developing cerebellum delineating its possible ontogenetic role. J. Neuroscience 33, 111-124.

Kuhn, T.B., Stoeckli, E.T., Condrau, M.A., Rathjen, F.G., & Sonderegger, P. (1991). Neurite outgrowth on immobilized Axonin-1 is mediated by a heterophilic interaction with L1 (G4). J. Cell Biol. 115, 1113-1126.

Lander, A.D., Stipp, C.S., & Ivins, J.K. (1996). The glypican family of heparan sulfate proteoglycans: Major cell-surface proteoglycans of the developing nervous system. In: *Proteoglycans in Neural Development* (R.U. Margolis & M. Grumet, Eds.).

Lemansky, P., Fatemi, S.H., Gorican, B., Meyale, S., Rossero, R., & Tartakoff, A.M. (1990). Dynanics and longevity of the glycolipid-anchored membrane protein, Thy-1. J. Cell Biol. 110, 1525-1531.

Lisanti, M.P., Caras, I.W., Davitz, M.A., & Rodriguez-Boulan, E. (1989a). A glycophospholipid membrane anchor acts as an apical targeting signal in polarized epithelial cells. J. Cell Biol. 109, 2145-2156.

Lisanti, M.P., Darnell, J.C., Chan, B.L., Rodriguez-Boulan, E., & Saltiel, A.R. (1989b). The distribution of glycosyl-phosphatidylinositol anchored proteins is differentially regulated by serum and insulin. Biochem. Biophys. Res. Commun. 164, 824-32.

Lisanti, M.P. & Rodriguez-Boulan, E. (1990). Glycophospholipid membrane anchoring provides clues to the mechanism of protein sorting in polarized epithelial cells. Trends Biochem. 15, 113-118.

Lisanti, M.P., Sargiacomo, M., Graeve, L., Saltiel, A.R., & Rodriguez-Boulan, E. (1988). Polarized apical distribution of glycosyl-phosphatidylinositol-anchored proteins in a renal epithelial cell line. Proc. Natl. Acad. Sci. USA 85, 9557-9561.

Low, M. (1989). The glycosyl-phosphatidylinositol anchor of membrane proteins. Biochim. Biophys. Acta. 988, 427-454.

Low, M.G. & Huang, K.S. (1991). Factors affecting the ability of glycosylphosphatidylinositol-specific phospholipase D to degrade the membrane anchors of cell surface proteins. Biochem J.

Luo, Y., Raible, D., & Raper, J.A. (1993). Collapsin: A protein in brain that induces the collapse and paralysis of neuronal growth cones. Cell 75, 217-227.

Mayford, M., Varzilai, A., Keller, F., Schacher, S., & Kandel, E.R. (1992). Modulation of an NCAM-related adhesion molecule with long-term synaptic plasticity in *Aplysia.* Science 256, 638-644.

Mayor, S., & Maxfield, F.R. (1995). Insolubility and redistribution of GPI-anchored proteins at the cell surface after detergent treatment. Mol. Biol. Cell 6, 929-944.

Mayor, S., Rothberg, K.G., & Maxfield, F.R. (1994). Sequestration of GPI-anchored proteins in caveolae triggered by cross-linking. Science 264, 1948-1951.

Metz, C.N., Brunner, G., Choi-Muira, N.H., Nguyen, H., Gabrilove, J., Caras, I.W., Altszuler, N., Rifkin, D.B., Wilson, E.L., & Davitz, M.A. (1994). Release of GPI-anchored membrane proteins by a cell-associated GPI-specific phospholipase D. EMBO J. 13, 1741-1751.

Metz, C.M., Thomas, P., & Davitz, M.A. (1992). Immunolocalization of a glycosylphosphatidylinsitol-specific phosphilipase D (GPI-PLD) in mast cells found in normal tissue and neurofibromatosis lesions. Amer. J. Path. 140, 1275-1281.

Metz, C.N., Zhang, Y.Y., Guo, Y., Tsang, T.C., Kochan, J.P., Altszuler, N., & Davitz, M.A. (1991). Production of the glycosylphosphatidylinositol-specific phospholipase D by the islets of Langerhans. J. Biol Chem. 266, 17733-17736.

Mikol, D.D., Gulcher, J.R., & Stefansson, K. (1990). The oligodendrocyte-myelin glycoprotein belongs to a distinct family of proteins and contains the HNK-1 carbohydrate. J Cell Biol. 110, 471-479.

Moss, D.J. & White, C.A. (1992). Solubility and posttranslational regulation of GP130/F11—a neuronal GPI-linked cell adhesion molecule enriched in the neuronal membrane skeleton. Eur. J. Cell Biol. 57, 59-65.

Nagafuchi, A. & Takeichi, M. (1989). Transmembrane control of cadherin-mediated cell adhesion: A 94 kDa protein functionally associated with a specific region of the cytoplasmic domain of E-cadherin. Cell Regulation 1, 37-44.

Noda, M., Yoon, K., Rodan, G.A., & Koppel, D.E. (1987). High lateral mobility of endogenous and transfected alkaline phosphatase: A phosphatidylinositol-anchored membrane protein. J. Cell Biol. 105, 1671-1677.

Olive, S., Dubois, C., Schachner, M., & Rougon, G. (1995). The F3 neuronal glycosylphosphatidylinositol-linked molecule is localized to glycolipid-enriched membrane subdomains and interacts with L1 and fyn kinase in cerebellum. J. Neurochem. 65, 2307-2317.

Ozawa, M., Baribault, H., & Kemler, R. (1989). The cytoplasmic domain of the cell adhesion molecule uvomorulin associates with three independent proteins structurally related in different species. EMBO J. 8, 1711-1717.

Parton, R.G., & Simons, K. (1995). Digging into caveolae. Science 269, 1398-1399.

Peles, E., Nativ, M., Campbell, P.L., Sakurai, T., Martinez, R., Lev, S., Clary, D.O., Schilling, J., Barnea, G., Plowman, G.D., Grumet, M., & Schlessinger, J. (1995). The carbonic anhydrase domain of receptor tryosine phosphatase beta is a functional ligand for the axonal cell recognition molecule contactin. Cell 82, 251-260.

Pesheva, P., Gennarini, G., Goridis, C., & Schachner, M. (1993). The F3/F11 cell adhesion molecule mediates the repulsion of neurons by the extracellular matrix glycoprotein J1-160/180. Neuron 10, 69-82.

Pimenta, A.F., Zhukareva, V., Reinoso, B., Grimley, C., Henzel, B., Fischer, I., & Leavitt, P. (1993). Cloning the limbic system-associated membrane protein (LAMP): A new immunoglobulin superfamily member. Soc Neurosci. Abstr. 19, 689a.

Pimenta, A.F., Zhukareva, V., Barbe, M.F., Reinoso, B.S., Grimley, C., Henzel, W., Fischer, I., & Levitt, P. (1995). The limbic system-associated membrane protein is an Ig superfamily member that mediates selective neuronal growth and axon targeting. Neuron 15, 287-297.

Powell, S.K., Cunningham, B.A., Edelman, G.M., & Rodriguez, B.E. (1991). Targeting of transmembrane and GPI-anchored forms of N-CAM to opposite domains of a polarized epithelial cell. Nature 353, 76-77.

Presky, D.H., Low, M.G., & Shevach, E.M. (1990). Role of phosphatidylinositol-anchored proteins in T cell activation. J. Immunol. 144, 860-868.

Prochiantz, A. (1995). Neuronal polarity: Giving neurons heads and tails. Neuron 15, 743-746.

Ranscht, B. & Dours-Zimmerman, M.T. (1991). T-cadherin, a novel cadherin cell adhesion molecule in the nervous system lacks the conserved cytoplasmic region. Neuron 7, 391-402.

Reichardt, L.F. & Tomaselli, K.J. (1991). Extracellular matrix molecules and their receptors: functions in neural development. Annu. Rev. Neurosci. 14, 531-570.

Roberts, J.M., Kenton, P., & Johnson, P.M. (1990). Growth factor-induced release of a glycosyl-phosphatidylinositol (GPI)-linked protein from the HEp-2 human carcinoma cell line. Febs Lett. 267, 213-216.

Robinson, P.J. (1991). Phosphatidylinositol membrane anchors and T-cell activation. Immunol. Today 12, 35-41.

Rosen, C.L., Lisanti, M.P. & Salzer, J.L. (1992). Expression of unique sets of GPI-linked proteins by different primary neurons *in vitro*. J. Cell Biol. 117, 617-627.

Rosse, W. (1990). Phosphatidylinositol-linked proteins and paroxysmal nocturnal hemoglobinuria. Blood 75, 1595-1601.

Sacristán, M.P., Vestal, D.J., Dours-Zimmerman, M.T., & Ranscht, B. (1993). T-cadherin 2: Molecular characterization, function in cell adhesion, and coexpression with T-cadherin and N-cadherin. J. Neurosci. Res. 34, 664-680.

Scallon, B.J., Fung, W.-J.C., Tsang, T.C., Li, S., Kado-Fong, H., Juang, K.-S., & Kochan, J.P. (1991). Primary structure and functional activity of a phosphatidylinositol-glycan-specific phospholipase D. Science 252, 446-448.

Schnitzer, J.E., McIntosh, D.P., Dvorak, A.M., Liu, J., & Oh, P. (1995). Separation of caveolae from associated microdomains of GPI-anchored proteins. Science 269, 1435-1439.

Schofield, P.R., McFarland, K.C., Hayflick, J.S., Wilcox, J.N., Cho, T.M., Roy, S., Lee, N.M., Loh, H.H., & Seeburg, P.H. (1989). Molecular characterization of a new immunoglobulin superfamily protein with potential roles in opioid binding and cell contact. EMBO J. 8, 489-495.

Seaman, W.E., Niemi, E.C., Stark, M.R., Goldfine, R.D., Pollock, A.S., & Imboden, J.B. (1991). Molecular cloning of gp42, a cell-surface molecule that is selectively induced on rat natural killer cells by interleukin 2: Glycolipid membrane anchoring and capacity for transmembrane signaling. J Exp. Med. 173, 251-60.

Seaver, E.C., Carpenter, E.M., & Bastiani, M.J. (1996). REGA-1 is a GPI-linked member of the immunoglobulin superfamily present on restricted regions of sheath cell processes in grasshopper. Development 122, 567-578.

Seeger, M.A., Haffley, L., & Kaufman, T.C. (1988). Characterization of *amalgam:* A member of the immunoglobulin gene superfamily from *Drosphila*. Cell 55, 589-600.

Sesko, A.M. & Low, M.G. (1991). Immunolocalization of glycosyl phosphatidyl-inositol-specific phospholipase D (GPI-PLD) in mammalian tissues. FASEB J. 5, A839.

Simons, K. & Wandinger-Ness, A. (1990). Polarized sorting in epithelia. Cell 62, 207-210.

Spaltmann, F., & Brümmendorf, T. (1996). CEPU-1, a novel immunoglobulin superfamily molecule, is expressed by developing cerebellar Purkinje cells. J. Neurosci. 16, 1770-1779.

Stahl, N., Boulton, T.G., Farruggella, T., Ip, N.Y., Davis, S., Witthuhn, B.A., Quelle, F.W., Silvennoinen, O., Barbieri, G., Pellegrini, S., Ihle, J.N., & Yancopoulos, G.D. (1994). Association and activation of Jak-Tyk kinases by CNTF-LIF-OSM-IL-6 β receptor components. Science 263, 92-95.

Stahl, N. & Prusiner, S.B. (1991). Prions and prion proteins. FASEB J. 5, 2799-2807.

Stefanová, I., Horeksí, V., Ansotegui, J., Knapp, W., & Stockinger, H. (1991). GPI-anchored cell-surface molecules complexed to protein tyrosine kinases. Science 254, 1016-1019.

Stipp, C.S., Litwack, E.D., & Lander, A.D. (1994). Cerebroglycan: An integral membrane heparan sulfate proteoglycan that is unique to the developing nervous system and expressed specifically during neuronal differentiation. J. Cell Biol. 124, 149-160.

Stoeckli, E.T., Kuhn, T.B., Duc, C.O., Ruegg, M.A., & Sonderegger, P. (1991). The axonally secreted protein axonin-1 is a potent substratum for neurite growth. J. Cell Biol. 112, 449-455.

Stoeckli, E.T., Lemkin, P.F., Kuhn, T.B., Ruegg, M.A., Heller, M., & Snderegger, P. (1989). Identification of proteins secreted from axons of embryonic dorsal root ganglia neurons. Eur. J. Biochem. 180, 249-258.

Struyk, A.F., Canoll, P., Wolfgang, M., Rosen, C.L., D'Eustachio, P., & Salzer, J.L. (1995). Molecular cloning of neurotrimin defines a new subfamily of differentially expressed neural cell adhesion molecules. J. Neurosci. 15, 2141-2156.

Su, G., Waneck, G.L., Flavell, R.A., & Bothwell, A.L. (1991). The glycosyl phosphatidylinositol anchor is critical for Ly-6A/ E-mediated T cell activation. J Cell Biol. 112, 377-384.

Takeda, J., & Kinoshita, T. (1995). GPI-anchor biosynthesis. Trends Biochem. Sci. 20, 367-371.

Takeda, J., Miyata, T., Kawagoe, K., Iida, Y., Endo, Y., Fujita, T., Takahashi, M., Kitani, T., & Kinoshita, T. (1993). Deficiency of the GPI anchor caused by a somatic mutation of the PIG-A gene in paroxysmal nocturnal hemoglobinuria. Cell 73, 703-711.

Tartakoff, A.M. & Singh, N. (1992). How to make a glycoinositol phospholipid anchor. Trends Biochem. Sci. 17, 470-473.

Thomas, P.M. & Samelson, L.E. (1992). The glycophosphatidylinsoitol-anchored Thy-1 molecule interacts with the $p60^{fyn}$ protein tyrosin kinase in T cells. J. Biol. Chem. 267, 12317-22.

Ting, A.E. & Pagano, R.E. (1990). Detection of a phosphatidylinositol-specific Phospholipase C at the surface of Swiss 3T3 cells and its potential role in the regulation of cell growth. J. Biol. Chem. 265, 5337-5340.

Tiveron, M.-C., Barboni, E., Rivero, F.B.P., Gormley, A.M., Seeley, P.J., Grosveld, F., & Morris, R. (1992). Selective inhibition of neurite outgrowth on mature astrocytes by Thy-1 glycoprotein. Nature 355, 745-748.

Toutant, J.-P., Richards, M.K., Krall, J.A., & Rosenberry, T.L. (1990). Molecular forms of acetylcholinesterase in two sublines of human erythroleukemia K562 cells: Sensitivity or resistance to phosphatidylinositol-specific phospholipase C and biosynthesis. Eur. J. Biochem. 187, 31-38.

Treanor, J.J.S., Goodman, L., de Sauvage, F., Stone, D.M., Poulsen, K.T., Beck, C.D., Gray, C., Armanini, M.P., Pollock, R.A., Hefti, F., Phillips, H.S., Goddard, A., Moore, M.W., Buj-Bello, A., Davies, A.M., Asai, N., Takahashi, M., Vandlen, R., Henderson, C.E., & Rosenthal, A. (1996). Characterization of a multicomponent receptor for GDNF. Nature 382, 80-83.

Tse, A.G., Barclay, A.N., Watts, A., & Williams, A.F. (1985). A glycophosphilipid tail at the carboxy terminus of the Thy-1 glycoprotein of neurons and thymocytes. Science 230, 1003-1008.

Vestal, D.J. & Ranscht, B. (1992). Glycosyl phosphatidylinositol-anchored T-cadherin mediates calcium dependent, homophilic cell adhesion. J. Cell Biol. 119, 451-461.

Walter, E.I., Roberts, W.L., Rosenberry, T.L., Ratnoff, W.D., & Medof, M.E. (1990). Structural basis for variations in the sensitivity of human decay accelerating factor to phosphatidylinositol-specific phosphilipase C cleavage. J Immunol. 144, 1030-1036. [Published erratum appears in J. Immunol. (1990, May 15), 144(10), 4072.]

Walter, J., Muller, B. & Bonhoeffer, F. (1990). Axonal guidance by an avoidance mechanism. J. Physiol. (*Paris*). 84, 104-110.

Williams, A.F., & Gagnon, J. (1982). Neuronal Thy-1 glycoprotein: Homology with immunoglobulin. Science 216, 696-703.

Xue, G.P., Rivero, B.P., & Morris, R.J. (1991). The surface glycoprotein Thy-1 is excluded from growing axons during development: A study of the expression of Thy-1 during axogenesis in hippocampus and hindbrain. Development 112, 161-176.

Yoshihara, Y., Kawasaki, M., Tamada, A., Nagata, S., Kagamiyama, H., & Mori, K. (1995). Overlapping and differential expression of BIG-2, BIG-1, TAG-1, and F3: Four members of an axon-associated cell adhesion molecule subgroup of the immunoglobulin superfamily. J. Neurobiol. 28, 51-69.

Yoshihara, Y., Kawasaki, M., Tani, A., Tamada, A., Nagata, S., Kagamiyama, H., & Mori, K. (1994). BIG-1: A new TAG-1/F3-related member of the immunoglobulin superfamily with neurite outgrowth-promoting activity. Neuron 13, 415-426.

Zhang, F., Crise, B., Su, B., Hou, Y., Rose, J.K., Bothwell, A., & Jacobson, K. (1991). Lateral diffusion of membrane-spanning and glycosylphosphatidylinositol-linked proteins: Toward establishing rules governing the lateral mobility of membrane proteins. J. Cell Biol. 15, 75-84.

Zinn, K., McAllister, L., & Goodman, C.S. (1988). Sequence analysis of neuronal expression of Fasciclin I in grasshopper and *Drosphila.* Cell 53, 577-587.

Zisch, A.H., D'Alessandri, L., Amrein, K., Ranscht, B., Winterhalter, K.H., & Vaughan, L. (1995). The glypiated neuronal cell adhesion molecule contactin/F11 complexes with src-family protein tyrosine kinase Fyn. Mol. Cell Neurosci. 6, 263-279.

Zuellig, R.A., Rader, C., Schroeder, A., Kalousek, M.B., Bohlen, F.v., Osterwalder, T., Inan, C., Stoeckli, E.T., Affolter, H.U., Fritz, A., Hafen, E., & Sonderegger, P. (1992). The axonally secreted cell adhesion molecule axonin-1: Primary structure, immunoglobulin-like and fibronectin-type III-like domains and glycosyl-phosphatidylinositol anchorage. Eur. J. Biochem. 204, 453-463.

INDEX